Building Beehives

FOR

DUMMIES®

Building Beehives

FOR

DUMMIES®

by Howland Blackiston

WILEY

John Wiley & Sons, Inc.

Building Beehives For Dummies®

Published by
John Wiley & Sons, Inc.
111 River St.
Hoboken, NJ 07030-5774
www.wiley.com

About the Author

Howland Blackiston has been a backyard beekeeper since 1984 and an artist and craftsman all his life. He's written hundreds of articles and has appeared on dozens of television and radio programs, including shows on the Discovery Channel, CNBC, CNN, NPR, Sirius XM Radio, and scores of regional shows. He has been a keynote speaker at conferences and events in more than 40 countries. Howland is the past president of Connecticut's Back Yard Beekeepers Association, one of the nation's largest regional clubs for the hobbyist beekeeper. Howland, his wife, Joy, and their bees live in Weston, Connecticut.

Howland is the author of *Beekeeping For Dummies*, 2nd Edition, published by John Wiley & Sons, Inc. You can reach Howland by e-mail. He'd love to hear from you: howland@buildingbeehives.com.

Dedication

I dedicate this book to my beautiful wife, Joy, who has patiently spent many evenings and weekends alone while I worked on this book. She is, and always will be, the queen bee of my universe.

Author's Acknowledgments

Writing this book has been a labor of love, thanks to the wonderful folks at John Wiley & Sons, Inc.: Senior Acquisitions Editor Tracy Boggier, who approached me in the first place about creating this companion piece to *Beekeeping For Dummies;* Senior Project Editor Georgette Beatty, whose sound advice and superb organizational skills kept me focused and on schedule; and Copy Editor Todd Lothery, whose supreme wordsmithing elevated my narrative into something to be proud of. Also many thanks to the composition team at Wiley, who handled nearly everything to do with the way the words and images ultimately appear on the page. Kudos to cartoonist Rich Tennant (I'm a big fan), who has the enviable ability to take any topic and find the humor in it.

Special thanks to my friend Michael Lund, who generously served as my technical consultant on this project. Michael is not only a gifted craftsman and woodworker but also a supremely knowledgeable and innovative beekeeper. That's a great combination when it comes to building beehives!

New York beekeeper Andrew Coté contributed his urban beekeeping wisdom to Chapter 2, Michael Paoletto graciously shared his plans for making a Kenya top bar hive (Chapter 5), Ellen Zampino provided the inspiration for the elevated hive stand (Chapter 13), and Dennis Murrell allowed me to adapt his plans for the four-frame observation hive (Chapter 7). Felix Freudzon, of Freudzon Design International, expertly created the 3-D instructional illustrations in the book.

Early on, as I was deciding what topics I should cover in this book, a number of the members of Connecticut's Back Yard Beekeepers Association participated in a focus group at my home. Their insights and suggestions were very helpful in homing in on the topics that would become the contents of the book. Many thanks to Jon Dickey, Joe Fischer, Jerry Goodwin, John Grimshaw, Leslie Huston, Gabriele Kallenborn, Michael Lund, Marina Marchese, Bobbie Meyzen, and Mike Paoletto.

Publisher's Acknowledgments

We're proud of this book; please send us your comments at http://dummies.custhelp.com. For other comments, please contact our Customer Care Department within the U.S. at 877-762-2974, outside the U.S. at 317-572-3993, or fax 317-572-4002.

Some of the people who helped bring this book to market include the following:

Acquisitions, Editorial, and Vertical Websites

Senior Project Editor: Georgette Beatty

Senior Acquisitions Editor: Tracy Boggier

Copy Editor: Todd Lothery

Assistant Editor: David Lutton

Editorial Program Coordinator: Joe Niesen

Technical Editor: Garry Reeves

Editorial Manager: Michelle Hacker

Editorial Assistant: Alexa Koschier

Cover Photo: © Howland Blackiston

Cartoons: Rich Tennant (www.the5thwave.com)

Composition Services

Project Coordinator: Katherine Crocker

Layout and Graphics: Jennifer Creasey, Joyce Haughey, Christin Swinford, Erin Zeltner

Proofreaders: John Greenough, Betty Kish

Indexer: Glassman Indexing Services

Illustrator: Felix Freudzon, Freudzon Design

Publishing and Editorial for Consumer Dummies

    Kathleen Nebenhaus, Vice President and Executive Publisher

    David Palmer, Associate Publisher

    Kristin Ferguson-Wagstaffe, Product Development Director

Publishing for Technology Dummies

    Andy Cummings, Vice President and Publisher

Composition Services

    Debbie Stailey, Director of Composition Services

Contents at a Glance

Contents at a Glance

Table of Contents

Introduction

• •

Beekeepers are a self-reliant lot. They're quick to find practical ways to care for their bees and discover innovative, cost-effective solutions to beekeeping problems. Typically, beekeepers are passionate about nature and the well-being of the environment. So it's no wonder that more and more bee-keepers are finding out how to build their own beehives and equipment. Sure, you can purchase kits from beekeeping supply vendors, but doing it yourself is very enjoyable and satisfying. You may even save some money! In today's green world of self-sufficiency, sustainability, and back-to-basics mentality, why not roll up your sleeves and make a beeline for a little side hobby that both you and your bees will appreciate and benefit from?

Whether you're a new beekeeper or a seasoned old-timer, this book provides you with a step-by-step approach for building some of the world's most popular hive designs and beekeeping accessories. And don't fret if you don't know much about carpentry or woodworking — I've kept the designs and instructions as simple and straightforward as possible.

About This Book

If you poke around the Internet you can drum up a variety of plans for building beehives and accessories. The problem is that almost none of these offers lucid, step-by-step instructions and illustrations for *how* to build and assemble these hives. None offers a detailed materials list so you know how much wood, hardware, and fasteners you need for a particular design. They often lack explanations of how certain designs benefit the bees and enhance your ability to be a better beekeeper. They seldom have instructions on how to make certain cuts or joinery.

In contrast, you'll find all the information and answers you need in this book: the pros and cons associated with each design, estimated building costs, degree of difficulty, detailed lists of what materials to buy, exact specs on how to cut the lumber, precise instructions and clear illustrations for assembly, and more. I even include a chapter on ten things you can do to add special touches and embellishments to the designs in this book (see Chapter 19).

This book includes designs for six different beehives (Part II) and seven beekeeping accessories (Part III), including frames. I selected the designs based on their overall popularity among today's backyard beekeepers and their compatibility with commercially available equipment, add-ons, and accessories.

I've organized the material in this book in a logical way to help you quickly find the information you need and go straight to it. Here's some good news for you skimmers out there: You don't have to read this book from cover to cover to build a beehive. If you're an old pro at woodworking, you can skip the text dealing with woodworking tools, equipment, materials, and skills. If you're an experienced beekeeper, you can skip the text that explains the components of a hive and their significance to bees. You're welcome to make a beeline to the design that beckons you and just start building! (But if you decide to read the book straight through and absorb all the information, I won't complain.)

Conventions Used in This Book

Before you get started, you should be aware of a few conventions — that is, standard formatting techniques that are used in this book:

- ✔ I use **bold** text to highlight the key words in bulleted lists and the action parts of numbered steps for assembling your hives and equipment.

- ✔ When I introduce a new term I put it in *italics* the first time and follow up with a simple definition. I also use italics to add emphasis.

- ✔ Website addresses appear in `monofont` to make them stand out.

- ✔ When this book was printed, some web addresses may have needed to break across two lines of text. If that happened, rest assured that I haven't put in any extra characters (such as hyphens) to indicate the break. So when using one of these web addresses, just type in exactly what you see in this book, pretending as though the line break doesn't exist.

- ✔ This book has mucho measurements and numerical notations. All measurements are in customary U.S. units (inches and feet). (Happily, you can find wonderful websites and smartphone apps to make the conversion from the U.S. system to the metric system if you need to.) In the tables that list materials, I express feet as a single quotation mark (as in 8') and inches as double quotation marks (as in 2"). And when I discuss board lumber, I use the letter *x* as an abbreviation for the word *by* (as in 2" x 3").

- ✔ When providing overall measurements, I always list them in this sequence: length x width x height (or, alternatively, thickness).

- ✔ Many of the plans in this book suggest using clear pine as a building material. By *clear* I mean pine lumber with straight, even grain and virtually no knots. Depending on where you live, clear pine may also be known as *select* pine or *grade C* pine. If you can't find this grade in your area, you can always use a lesser grade with knots (the bees won't care), but you may have to plan on some extra material in case a pesky knot winds up being right where you planned to join two pieces together.

- ✔ When you purchase lumber at the store, you order it based on what's known as its *nominal* size. But here's the confusing part — the actual size of the lumber is always a little bit smaller. *Nominal* refers to its rough dimension, before it's trimmed and sanded to its finished size at the lumber mill. When a two-by-four is cut out of a log it is in fact 2 inches by 4 inches. But after the board is dried and planed it becomes its *actual* size of 1½ inches by 3½ inches. In the chapters, the materials lists refer to the nominal sizes of lumber, and the cut sheets refer to the actual size.

What You're Not to Read

You find information that's interesting but not essential to your understanding of building beehives in two spots: sidebars (shaded gray boxes) and text that's designated with the Technical Stuff icon. You don't have to read this info, but I'll be happy if you do!

Foolish Assumptions

In tailoring this book for you, I've made a few assumptions:

- You're a beekeeper or hope to become a beekeeper, and you're interested in building your own beehives, accessories, and equipment.

- You already know enough about beekeeping that I don't have to explain how to care for your bees. If you want to learn more about beekeeping, check out my other book, *Beekeeping For Dummies,* 2nd Edition (Wiley).

- You're a fairly competent *DIYer* (do-it-yourselfer) or you're interested in discovering some basic skills that allow you to build your own beehives and accessories. I didn't write this book to show you woodworking skills (although I cover some basics in Chapter 4). I wrote this book for the neophyte or advanced woodworker who's eager to build some beehives and accessories. You needn't be an expert, and you needn't have a workshop full of fancy, expensive woodworking equipment. But the spirit is willing. If you can drive a nail and cut a straightedge with a table saw, you'll be completely comfortable building any of the designs in this book. I've graded the various designs based on their degree of building difficulty. If you're a little shaky at this carpentry thing, start with an easy build (to hone your skills) before tackling plans that are a bit more complex.

- You know how to read schematic plans and how to build from them (I also include step-by-step narrative assembly instructions to round out the information you glean from the drawings).

How This Book Is Organized

The book features four parts, with several chapters in each part. Each chapter is broken down into smaller, more digestible sections that you can easily identify by headings in bold type. I also include lots of photos and illustrations (each, I hope, is worth a thousand words). The following sections describe how the book is structured.

Part 1: The Buzz on Beehives

Chapter 1 provides a brief overview of all the major topics in this book. In Chapter 2 you find out more about the remarkable honeybee and discover the hive features that are most critical for the bees' shelter, safety, and ability to grow and prosper. Also in Chapter 2, I help you figure out the best location for your hives and show you how to satisfy the needs of neighbors and abide by regional laws and regulations. And I include info that helps you decide which of the hive designs in this book best meets the needs of your bees and you, the beekeeper.

Chapter 3 is devoted to identifying the tools and equipment you need to build the hives and accessories in this book. And Chapter 4 has an array of helpful tips and instructions for fine-tuning basic carpentry skills.

Part II: The World's Most Popular Beehive Designs

This part provides detailed materials and cut lists, plus step-by-step illustrated instructions, for building the following popular hive designs (listed here based on degree of difficulty to build, starting with the easiest first):

- ✔ Kenya top bar hive and stand
- ✔ Five-frame nuc hive
- ✔ Four-frame observation hive
- ✔ Warré hive
- ✔ British National hive
- ✔ Langstroth hive (both eight- and ten-frame models)

Part III: Sweet Beehive Accessories

This part provides detailed materials and cut lists, plus step-by-step illustrated instructions, for building and assembling the following beekeeping accessories:

- ✔ Frame jig (to help you assemble frames quickly and easily)
- ✔ Double screened inner cover (for the eight-frame Langstroth hive, the ten-frame Langstroth hive, the nuc hive, and the British National hive)
- ✔ Elevated hive stand
- ✔ IPM screened bottom board
- ✔ Hive-top feeder (for the eight-frame and ten-frame Langstroth hive)
- ✔ Solar wax melter
- ✔ Langstroth frames (shallow, medium, and deep)

Part IV: The Part of Tens

Finally, no *For Dummies* book is complete without the Part of Tens, so I offer a collection of extra take-away information not found elsewhere in the book.

In Chapter 18 I share ten things you can do to better maintain your hives and accessories to give you added years of service. Chapter 19 presents ten fun ideas for adding extra features, doodads, and options that will make your hives unique and bling-o-rific.

Icons Used in This Book

Peppered throughout this book are helpful icons that present special types of information to enhance your reading experience.

I use this icon to point out things that need to be so ingrained in your consciousness that they become habits. Keep these points at the forefront of your mind when building your hives and accessories.

This icon highlights information that's interesting (to me, at least) but not crucial to your understanding of building beehives.

Think of these tips as words of wisdom that — when applied — can make your experience building hives and accessories more pleasant and fulfilling!

These warnings alert you to potential boo-boos that can make your beehive-building experiences unpleasant or even dangerous. Take them to heart!

Where to Go from Here

This book isn't linear, meaning you don't have to read everything sequentially from start to finish. Feel free to buzz around as your needs and interests dictate. But here are a few helpful hints:

- If you want to find out how the hives you build will support your bees' needs, start with Chapter 2. This chapter helps you understand how a colony uses a hive, helps you decide where to locate your hives, and guides you in choosing a hive design that best matches your beekeeping objectives and your woodworking skills.

- If you're ready to start thinking about what tools and materials you need to build your hives and accessories, flip to Chapter 3.

- Feeling a little uncertain about your woodworking skills? Chapter 4 can help you brush up on some good, basic carpentry techniques that you'll use throughout the builds.

- If you're ready to get on with it and check out some building plans, rush right to Parts II and III.

Happy building!

Part I
The Buzz on Beehives

The 5th Wave   By Rich Tennant

"Let me do the talking."

In this part . . .

Here's where I set the groundwork for finding out the basics of building beehives. First I cover the structure of a beehive, and then I tell you how to select the best hive for your needs and skills, where to place your hives, and what tools and materials you need to build them. I even give you some suggestions for fine-tuning your carpentry skills. This part has all the info you need to get ready for your adventure in building beehives and beekeeping equipment.

Chapter 1

Getting Sweet on Building Your Own Beehives

. . . .

In This Chapter

▶ Getting the scoop on bees and their lives

▶ Seeing the advantages of building (versus buying) hives

▶ Planning your beehive build

▶ Putting together the right workspace, tools, and materials

▶ Honing your woodworking skills

▶ Digging into beehive designs

. . . .

My backyard beekeeping adventure started in 1983, and I've never ceased to be amazed by these endearing creatures — the profound contribution they bring to gardening and agriculture through their pollination services; their remarkable social and communication skills; and, of course, that wonderful bonus of a yearly harvest of pure, all-natural, delicious honey. It's no wonder that beekeepers speak with such warmth about their *girls*.

As a beekeeper, it doesn't take long to expand the scope of your hobby into related beekeeping adventures, such as candle making, mead brewing, and a host of other honeybee-related activities. And if you have even a remote interest in woodworking and building things, it's only natural to want to learn how to build a home for your beloved bees.

In this chapter I include some information to help you get ready for building your own beehives and accessories. I start with Honeybees 101 — a mini review of the bountiful bee and what goes on inside a beehive. Then I turn to some ideas for setting up your beehive-making shop, fine-tuning your woodworking skills, and deciding which of the plans in this book best meet your needs and skills.

Bee-ing in the Know about Bees

So you want to build some hives for your precious bees? You're going to have fun! You have many options regarding the hives you can build, but before you dig in, it's helpful to understand a little bit about these extraordinary creatures.

Why honeybees are great pollinators

Honeybees can outperform all other types of pollinators in nature for a number of reasons.

✔ The honeybee's body and legs are covered with branched hairs that effectively catch and hold pollen grains. When a bee brushes against the stigma (female part) of the next flower she visits, some of the pollen grains from the previous flower are deposited, and the act of cross-pollination is accomplished.

✔ Unlike other pollinating insects that lie dormant all winter and then emerge in the spring in very small numbers, the honeybee colony overwinters, with thousands of bees feeding on stored honey. Early in the spring, the queen begins laying eggs, and the already large population explodes to many tens of thousands of bees that carry out pollination activities.

✔ The honeybee tends to forage on blooms of the same kind, as long as they're flowering, versus hopping from one flower type to another. This single focus makes for particularly effective pollination.

✔ The honeybee is one of the few pollinating insects that can be introduced to a garden at the gardener's will.

Honeybees' most important job: Pollination

Honeybees are a critical part of the agricultural economy. They account for more than 80 percent of all pollination of crops. In fact, honeybees pollinate more than 100 cultivated crops, including various fruits and vegetables, nuts, herbs, spices, and numerous ornamental plants. According to the agriculture department at the University of Arkansas, honeybees add an estimated $15 billion to the U.S. economy each year in increased crop yields.

Since 2006, the population of honeybees has been dwindling at an alarming rate. The reasons for this die-off of colonies are not yet fully understood at the time of this writing. But the consequence is laser-sharp. A spring without bees would seriously endanger our food supply. Building hives and establishing some colonies of bees in your neighborhood makes an important contribution to reintroducing pollinating bees to your neck of the woods.

The products of the honeybee

In addition to the wonderful pollination services that honeybees provide (see the preceding section), they produce products that you can harvest and put to all kinds of uses. These products include

✔ **Beeswax:** Honeybees secrete wax from eight glands located along their abdomen. They use beeswax to build the hexagonal cells in which they raise their brood and store their honey and pollen. You'll probably get several pounds of surplus wax for every 100 pounds of honey that you harvest. You can clean and melt down this wax for all kinds of uses, including candles, furniture polish, and cosmetics. Pound for pound, wax is worth more than honey, so it's definitely worth a bit of effort to reclaim this prize.

In this book, the Kenya top bar hive and the Warré hive (see Chapters 5 and 8, respectively) give you a lot of beautiful wax because, with these particular hives, you remove and crush the honeycomb to harvest your honey. To render the wax you collect from your hives, use a solar wax melter (see Chapter 16 for instructions on how to build one).

✔ **Honey:** Bees use honey as food, just like humans do. It's their carbohydrate. For people, eating local honey is said to relieve the symptoms of pollen-related allergies.

There's something magical about bottling your own honey, and I can assure you that no other honey tastes as good as the honey made by your own bees. How much honey can you expect? The answer to that question varies depending on the weather, rainfall, and location and strength of your colony. But producing 40 to 80 pounds or more of surplus honey per hive isn't unusual.

In this book, the best hives for producing copious amounts of the precious liquid gold include (in potentially dwindling order of abundance) the Langstroth hive (see Chapter 10), the British National hive (see Chapter 9), the Warré hive (see Chapter 8), and the Kenya top bar hive (see Chapter 5).

✔ **Pollen:** Bees use pollen like they use honey — as food. And why not? Pollen is one of the richest and purest of natural foods, consisting of up to 35 percent protein and 10 percent sugars, carbohydrates, enzymes, minerals, and vitamins including A (carotenes), B1 (thiamin), B2 (riboflavin), B3 (nicotinic acid), B5 (panothenic acid), C (ascorbic acid), and H (biotin).

You can harvest pollen from your bees using a pollen trap (they're available from any beekeeping supply house). You can sprinkle a small amount on your breakfast cereal or in yogurt (as you might do with wheat germ). I like to sprinkle some on salads as a colorful addition. It's said that eating a little local pollen every day can relieve the symptoms of pollen-related allergies. When you have your own beehives, all-natural allergy relief is only a nibble away! Both the British National hive (Chapter 9) and the Langstroth hive (Chapter 10) lend themselves to effective pollen harvesting, as these are the hive types for which commercially made pollen traps are available.

✔ **Propolis:** Sometimes called *bee glue*, this super-sticky, gooey material is gathered by bees from trees and plants. The bees use this brown goop to fill drafty cracks in the hive, strengthen comb, and sterilize their home. Propolis contains vitamins, minerals, amino acids, and flavonoids that are said to promote anti-inflammatory, antioxidant, and antibiotic properties. You'll see a number of products in health food stores that contain propolis — everything from toothpaste to emollients to cough drops. You can harvest propolis from any of the hives in this book by simply scraping it off of the hive surfaces with your hive tool. In addition, many beekeeping supply vendors sell special propolis traps that encourage a particularly large harvest of the goo. Propolis can be rendered at home into various products, including medicinal tinctures, and even a top quality wood varnish.

Propolis has remarkable antimicrobial qualities that guard against bacteria and fungi. Its use by bees makes the hive one of the most hygienic domiciles found in nature. This property hasn't gone unnoticed over the centuries. The Chinese have used it in medicine for thousands of years. Even Hippocrates touted the value of propolis for healing wounds.

✔ **Royal jelly:** Royal jelly is a creamy substance made of digested pollen and honey or nectar mixed with a chemical secreted from a gland in a nurse bee's head. It transforms an ordinary worker bee into a queen bee and extends her life span from six weeks to five years!

In health food stores, royal jelly commands premium prices rivaling imported caviar. Products containing royal jelly are sold as dietary supplements that boast all kinds of benefits, including weight control, energy stimulant, skin health, and even improved reproductive health and fertility. Royal jelly contains an abundance of nutrients, including essential minerals, B-complex vitamins, proteins, amino acids, collagen, and essential fatty acids, just to name a few. Using an eyedropper or an itty-bitty spoon designed for this purpose, you can harvest royal jelly from the queen cells in your hives (that's the primary place the bees deposit it) and sell it for a pretty penny.

Any of the hives in this book (except for the four-frame observation hive in Chapter 7) would provide you with this opportunity. The larger the hive, the larger the harvest. But note that a large number of colonies are required to harvest anything beyond a minimal amount of royal jelly.

Although the health benefits of ingesting honey, pollen, propolis, and royal jelly have been touted for centuries, keep in mind that there's a relatively small percentage of the population that can have a severe and dangerous allergic reaction to ingesting the products of the honeybee. If you don't know your own situation, play it safe and consult your doctor or allergist before adding these products to your diet.

The main players in a beehive

In nature, honeybees typically build their hives in the hollow of a tree or some other cave-like environment. They like to be off the ground, safe from predators, and well protected from harsh weather. The hives in this book emulate, to varying degrees, the conditions bees prefer in nature while providing you (the beekeeper) with features that allow for easy inspections and manipulations to encourage strong and healthy colonies.

So what actually goes on inside a beehive? The following sections note the three main types of bees in a hive and what they do.

Her majesty, the queen

The *queen bee* is the heart and soul of the colony. She's the reason for nearly everything the rest of the colony does, and without her, the colony wouldn't survive. Only one queen lives in a given hive. A good-quality queen results in a strong and productive hive.

The queen's purpose is to lay eggs — lots of them. She is, in fact, an egg-laying machine, capable of producing more than 1,500 eggs a day at 30-second intervals. As a beekeeper, one of your primary objectives when inspecting your colonies is to confirm that each colony has a queen and that she's doing a good job of laying eggs and raising healthy brood. That's why many of the hives in this book include design features that make such inspections easy for the beekeeper.

Industrious worker bees

During the active season, more than 90 percent of the colony's population consists of *worker bees.* Like the queen, worker bees are female, but these girls are unable to mate and lack fully developed ovaries.

The term "busy as a bee" is well earned. Worker bees do a lot of work. They do it tirelessly, day in and day out. From the moment a worker bee is hatched, she has many and varied tasks clearly cut out for her. As she ages, she performs more complex and demanding tasks. Although these various duties usually follow a set pattern and timeline, they sometimes overlap. Worker bees may feed and care for the brood (nurse bees), feed and care for the queen (royal attendants), build comb, clean and do general housekeeping, guard the hive, and, of course, forage for pollen and nectar.

The more worker bees you have, the better it is for the colony, the more effective the colony's pollination services, and the greater the honey production. A number of the hive designs in this book allow for virtually unlimited growth of the colony. The Warré hive, the British National hive, and the Langstroth hive (see Chapters 8, 9, and 10 respectively) all include a modular design to provide such expansion.

Woeful drones

Drones are the only male bees in a colony, and they make up a small percentage of the hive's total population (less than 10 percent at the height of the season).

Procreation is the drone's primary purpose in life. Despite their high maintenance (the worker bees must feed and care for them), drones are tolerated and allowed to remain in the hive because they may be needed to mate with a new virgin queen (when the old queen dies or needs to be superseded).

When the weather gets cooler and the mating season comes to a close, the worker bees don't tolerate having drones around. After all, those fellows have big appetites and would consume a tremendous amount of food during the perilous winter months. So at the end of the nectar-producing season, the worker bees systematically expel the drones from the hive. They are literally tossed out the door.

Any of the hives in this book allow you to witness the banishment of drones. This is your signal that the beekeeping season is over for the year. Time to get back in the workshop and build some hives and equipment as holiday gifts!

Appreciating the Benefits of Building a Beehive Rather than Buying One

Many beekeeping supply companies are out there; a web search turns up dozens of them. Although you can certainly purchase a ready-made kit that will result in a wonderful home for your bees, building hives from scratch makes sense for all kinds of reasons. The following sections list some of the motivations that may prompt you to do it yourself, wind up with a potentially better product, and have more fun while you're at it.

Have fun and feel self-satisfaction

Building your own hives can be a wonderful new hobby that you can feel proud of. Knowing that the hives and equipment you use were made by your own hands is very satisfying. And the woodworking itself is enjoyable, from the intoxicating scent of the fresh-cut wood to the pleasure that comes with any do-it-yourself project.

I have another DIY hobby — I tap the sugar maple trees on my property and boil the sap down into delicious maple syrup. It's a hugely time-consuming project. My wife keeps reminding me that I could purchase maple syrup in the store for a fraction of the cost and for far less work. That may be true, but the fulfillment I feel from doing this myself far outstrips my wife's practicality.

In a way, the same is true with building beehives. Sure, you can buy a kit, and that may be cheaper than building your own. But making a hive from scratch is just plain fun and worth the effort. Plus, I'm convinced that the hives I build myself are better in every way than a store-bought hive.

Enhance your commitment to beekeeping, and better understand your bees' home

Building and better understanding beehives and equipment provide you with a greater awareness of what's most important to your honeybees and why. Rather than just reading about hives in a book, you can witness firsthand how the various features of different hive designs impact the productivity of your colonies and your ability to be a more effective beekeeper.

Modify designs to better meet your needs

The designs in this book are tried and true, but they're not necessarily the bee-all and end-all of hive designs. By all means, experiment and try some tweaks and modifications to what I suggest. You may come up with something that makes your life easier as a beekeeper or results in an improved all-around design for your bees. If you think you've stumbled upon a real breakthrough improvement, I'd love to hear about it — and so would others, I dare say. Be sure to drop me a line at howland@buildingbeehives.com.

Enjoy better quality than store-bought kits

Don't get me wrong, a lot of commercial beehive kits are very nice. But when you build hives yourself, you can add that certain *je ne sais quoi* that sets your hives apart from store-bought — that extra effort to achieve better fitting parts, slight tweaks in design to better suit your needs, or the use of nicer lumber and fasteners than come with mail-order kits. You know the old saying, *If you want it done right, do it yourself.*

Go green and recycle

If you have access to scrap wood, you're in an enviable position. The materials costs for building any of the hives or equipment in this book will be next to nothing. And besides the financial savings, you'll be doing the right thing by joining the admirable ranks of sustainability, recycling, and saving the environment. Congratulations!

If using scrap wood, be certain that it's free of any chemicals. You don't want to introduce any toxins to your bees.

Make building a family affair

Like beekeeping itself, building hives and equipment can be a fun family project. Now I don't suggest turning over table saw duties to your 10-year-old, but kids will have ample opportunities to help sand, paint, assemble, and accessorize your new hives and equipment. Or, perhaps like Tom Sawyer, you may slyly persuade others of the great privilege associated with doing those tasks you'd rather not do yourself.

Sell your handiwork

Building beehives is a business opportunity. Some of the hives in this book (such as the Kenya top bar hive in Chapter 5 and the Warré hive in Chapter 8) are gaining popularity among backyard beekeepers, but these hives aren't usually available from many of the major equipment suppliers. Beekeepers in your neighborhood may swarm to someone who builds quality hives and accessories. Put your newly found talents to good use and start up a side-line business for yourself. Your local bee club is a great place to start marketing your goods. And don't forget, a handmade hive makes a stunning gift for that special beekeeper in your life.

Making Plans for Your Own Beehive

This book has six beehive designs to choose from, as well as seven accessory designs. Perhaps you'll build them all, or perhaps you'd like a little help deciding which are the best for your situation. Chapter 2 provides some helpful guidelines for selecting the plans that best meet your needs (and match your skills).

Before you start building stuff, it's helpful to have a better understanding of where to locate your hives, how to deal with neighbors, and what you need to know about local laws and ordinances regarding beekeeping. I include these and other considerations in Chapter 2. I also cover the bees' basic needs regarding shelter and safety, and I explain the function and purpose of each of the basic components of today's most widely used hive, the Langstroth hive.

Setting Up Your Workshop

You need a place to build your hives and equipment. It may be a basement, a garage, or a corner of your apartment, or perhaps you're lucky enough to already have a dedicated workshop space for woodworking projects. In any event, the power tools, supplies, and inventory take up a fair amount of space and create quite a mess (lots and lots of sawdust). If you can, latch on to a location with at least 150 to 250 square feet of space. That should be ample for safely working on the projects in this book. It's a bit less than the size of a single bay in a garage.

Here are some basic requirements to consider:

- ✔ **Let there be light:** If you wind up in a basement or garage, the lighting may be rather underwhelming. In this case, get some shop lights (long, fluorescent or LED bulbs in a simple fixture) to supplement what's there now. The more light on these projects the better. You'll be working with some potentially dangerous pieces of equipment (power drills and saws), and you certainly want to be able to clearly see what you're doing. Brighter is better.

- ✔ **The buzz on electrical needs:** A number of the tools you'll be using are electrical, so your workspace needs ample outlets. The good news is that most consumer-grade tools run on normal household current. But some of these tools (like table saws) draw a lot of power (amperage), so be sure your workshop outlets have enough amps to run the equipment you intend to use. Otherwise, you'll be busy as a bee running back and forth to the circuit breaker.

- ✔ **Keeping it safe:** Working with power equipment and sharp, wood-cutting tools is inherently dangerous. You can do some real damage to yourself if you don't take safety precautions. Chapter 3 gives you the lowdown on creating a safe environment in which to build your hives and equipment.

Assembling Tools and Materials

If this were a book on building a house or making custom cabinetry, the list of tools would be very, very long. But hey, you're only building beehives and some related equipment. These plans are mostly variations on building a box, so I've intentionally kept the list of required tools and fasteners to an absolute minimum. I'll bet you already have some of these tools in a drawer somewhere (hammer, screwdriver, and tape measure). Chapter 3 gives you the skinny on the basic tools you'll use for the projects in this book.

The list of stuff you need for building beehives doesn't end with tools, though; you need actual building materials, too. In this book, I try to recommend materials that are most cost-effective and readily available. Pine is always a great bet — readily available, easy to work with, and cost-effective. But you need not limit yourself to the ordinary. Have fun and experiment using different woods, fasteners, and hardware that go beyond the mundane. Chapter 3 has ideas for various building materials and hardware, and Chapter 19 gives you ten ideas for adding beaucoup bling to your hives.

Getting a Handle on Carpentry Skills

I devote Chapter 4 to the skills you need before you jump into your beehive-making adventure. They include the following:

- ✔ Understanding *bee space* (the crawl space needed by a bee to pass easily between two structures)
- ✔ Measuring and marking materials
- ✔ Cutting lumber
- ✔ Building different kinds of joints
- ✔ Working with tin and wire
- ✔ Assembling a hive with a square, nails, screws, and glue

Having a handle on carpentry skills makes life a lot easier for the bees (by providing them with ideal living conditions) and makes the construction of the hive and accessories much easier for you.

Constructing Hives and Accessories

Beehives come in many styles, but obviously, in writing this book, I was limited to what I could fit between the covers. So with a little help from my backyard beekeeping colleagues, I identified the hives that would likely be the most popular with the widest range of hobbyists. The book has detailed materials lists, cut sheets, and assembly instructions and drawings for building the following hives (the ones listed first are the easiest to build):

- ✔ **Kenya top bar hive** (Chapter 5): One of the oldest beehive designs, this simple-to-build and cost-effective hive is gaining new popularity among backyard beekeepers. You won't find these for sale from any of the major beekeeping supply stores.
- ✔ **Five-frame nuc hive** (Chapter 6): Every backyard beekeeper can benefit from having a nuc on hand. They're handy for housing a captured swarm, serving as a nursery for raising a queen, or providing a small, easy-to-manage colony of pollinating bees in your garden.
- ✔ **Four-frame observation hive** (Chapter 7): This observation hive is big enough to support a small colony throughout the season yet small enough to serve as a portable teaching tool. It also provides a safe, up-close view of bee behavior, giving you a visual barometer of what's likely happening in your full-size hives.
- ✔ **Warré hive** (Chapter 8): This efficient design uses top bars (rather than frames and foundation) and has been gaining renewed popularity among beekeepers seeking more natural approaches to beekeeping. The design provides a living arrangement that's similar to how bees live in the wild. You won't find these for sale from any of the major beekeeping supply stores.
- ✔ **British National hive** (Chapter 9): This attractive hive is very popular in the United Kingdom, and I include it here for that reason. It's a little smaller than the Langstroth hive, and therefore slightly lighter in weight.

✔ **Langstroth hive** (Chapter 10): Without any doubt, the Langstroth hive is the most popular and widely used hive today. Certainly this is the case in the United States and in most developed countries around the world. You can't go wrong with this design. It's great for pollination, great for honey production, and easy to inspect. I've included designs for both eight- and ten-frame models of this hive.

The book also includes designs for popular and practical accessories and add-ons for your beehives. They, too, are listed from easiest to build to hardest. Accessories include the following:

✔ **Frame jig** (Chapter 11): This is a nifty gadget that greatly simplifies and speeds the task of assembling Langstroth-style frames and foundation.

✔ **Double screened inner cover** (Chapter 12): For the ultimate in hive ventilation, you'll love this double screened inner cover. It's used in place of a conventional inner cover and features an access that you can open or close depending on the orientation of the inner cover. I've included plans for eight- and ten-frame Langstroth hives, as well as the five-frame nuc hive and the British National hive.

✔ **Elevated hive stand** (Chapter 13): Getting hives up and off the wet ground improves circulation and helps prevent your hives from rotting out. This particular design can accommodate the nuc, British National, or Langstroth hive.

✔ **IPM screened bottom board** (Chapter 14): A screened bottom board has become a standard part of Integrated Pest Management (IPM). It's used to improve ventilation and to help monitor and manage varroa mite infestations. This design fits a ten-frame Langstroth hive, but you can adjust its dimensions to fit other hives in the book.

✔ **Hive-top feeder** (Chapter 15): Most beekeepers need to feed their bees at one time or another. Here's a great way to provide syrup to the colony without having to smoke and disturb the bees. I include feeder plans for both the eight- and ten-frame versions of the Langstroth hive.

✔ **Solar wax melter** (Chapter 16): Beeswax commands a higher price per pound than honey. You can harvest and render it for selling or making cosmetics and candles. This design uses the free power of the sun as its energy source.

✔ **Langstroth-style frames** (Chapter 17): Frames are certainly one of the more challenging accessories you can choose to build. Though most readers will be content buying frames for their Langstroth hives, for the adventuresome and more experienced woodworkers, this chapter provides detailed plans and how-to illustrations. I include plans for deep, medium, and shallow frames. Have fun!

Don't limit yourself to the prescriptions in this book. Everything I recommend is intended to keep the process of building beehives as simple and cost-effective as possible, but feel free to stray from my suggestions to build something truly unique and wonderful. It's totally up to you. Chapter 19 includes ten ideas for improving upon and jazzing up your hives and equipment. Get creative with accessories, try different add-ons and finishes, or use a more aesthetic joinery technique. All this won't make much difference to the bees, but it will surely be satisfying to you and provide immeasurable bragging rights.

Having worked hard on building this stuff, you'll certainly want your handi-work to last as long as possible. So in Chapter 18 I provide ten tips for ensur-ing that your hives and equipment will last many, many seasons.

Chapter 2

Comb Sweet Comb: Beehive Basics

. . . .

In This Chapter

▶ Understanding the bees' bare necessities

▶ Checking out the basic parts of (nearly) all hives

▶ Making locale considerations

▶ Building a hive that meets your needs

. . . .

You'll soon be embarking on a wonderful hive-building adventure. You have many different hives from which to choose, but before deciding what to build, understanding the basic components of a typical beehive is helpful. You need to consider what's important to the bees, what's important to your community, what's important to you as a beekeeper, and what's manageable to you as a woodworker. This chapter gives you the scoop.

Knowing What Bees Need in a Hive

The hive is the bees' home. As doting beekeepers, we aim to provide an environment for our bees that meets or exceeds the needs they seek out in nature. Here are the basic housing requirements for raising happy, healthy, and productive colonies.

Shelter and safety

In the wild, honeybees don't nest underground (like bumblebees and yellow jackets). Instead, honeybees seek shelter and safety *above* ground, typically in the spacious hollow of an old tree. This arrangement keeps the colony much drier than the underground alternative and provides much more room and ventilation than cramped tunnels drilled into the soil. Anything approaching the bees' feral conditions is very attractive to the colony.

The hives in Part II are designed to mimic honeybees' preferred conditions in the wild. These man-made hives simulate and even improve upon the "cavity" bees seek for shelter and safety in the wild. (Bonus: Man-made hives also provide a degree of convenience for the beekeeper, allowing for easy inspections and manipulations to encourage large, healthy colonies and a bountiful honey harvest.)

The ability to expand

Colonies of honeybees grow in population, both in the wild and in a man-made hive. And that's a good thing. Larger colonies have a greater chance of survival by collecting more food. And a large population of bees means more warmth during the cold winter. Beekeepers want to encourage this growth. The more bees in a hive, the more bees you have for pollination and honey production. If you don't give a flourishing colony enough room, however, they'll likely swarm, effectively cutting the size of your colony in half. In that case, your colony's productivity is seriously compromised. So if you're keeping bees to optimize pollination and honey production, you want to build hives that address this expansion issue. Most of the so-called "modern" hives (such as the Langstroth, Warré, and British National featured in Part II) can easily expand in size as the colony grows.

Dry and well-ventilated conditions

Honeybees do remarkably well in all kinds of extreme weather conditions, provided you keep the hive dry and well ventilated. Excess moisture and the inability of the bees to regulate the hive's temperature can spell deep trouble for the colony. Maintaining ideal conditions can sometimes be a challenge for the bees in the wild, but here's where the beekeeper can help. When building your hives, provide good airflow. Screened inner covers (Chapter 12), elevated hive stands (Chapter 13), and screened bottom boards (Chapter 14) all help the bees maintain the ideal living conditions.

A nearby source of water

During their foraging season, bees collect more than just nectar and pollen. They gather a whole lot of water. They don't use it primarily to quench their thirst; they use it to dilute honey that's too thick and to cool the hive during hot weather. Field bees bring water back to the hive and deposit it in cells, while other bees fan their wings furiously to evaporate the water and regulate the hive's temperature. They look for a source of water that's nearest to the hive (they're practical, not lazy).

If your hive is at the edge of a stream or pond, that's perfect. If it isn't, you should provide a *nearby* water source for the bees. There are a couple of reasons for this. For one, you don't want your bees wasting a lot of energy traveling long distances to fetch water. That energy is better used collecting nectar. And two, remember that the bees will seek out the *nearest* water source, and you certainly don't want that source to be your neighbor's kiddie pool. Or, if you're in an urban setting, you want to keep the bees away from your neighbor's air conditioner drips. So what should you do? You can place a watering device closer to the hives than the alternative source. I put my watering devices right next to my hives.

You can improvise all kinds of watering devices: a hive-top feeder filled with water rather than syrup, a pie pan filled with gravel and topped off with water, a chicken-watering device (available at farm supply stores), or simply an outdoor faucet that's encouraged to develop a slow drip.

Here's a nifty idea that I learned from a fellow beekeeper. Find or purchase a clean pail or bucket. Any size, color, or material will do. Just make sure that it's clean and has never been used for chemicals, fertilizers, or pesticides. Drill a few ½ inch drainage holes around the top edge of the bucket. Four or so should do the trick. Place the holes about 1 to 3 inches down from the top edge. Fill the bucket nearly to the holes with water, and then float a single layer of Styrofoam packaging pellets on the surface (see Figure 2-1). The pellets give the bees something to stand on as they sip water. That way they won't drown. Alternatively, you could use wood chips, sticks, or any other floating platform the bees can get a good footing on. The holes around the top are drainage holes that keep rainwater from overflowing the bucket and washing away the pellets, sticks, or wood chips. Neat, huh?

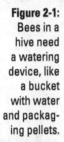

Figure 2-1: Bees in a hive need a watering device, like a bucket with water and packaging pellets.

Photograph courtesy of Howland Blackiston

Understanding the Anatomy of a Beehive

One way or another, made-made beehives are designed to provide the bees with shelter from the elements, a space to raise brood, a space to store honey, and adequate ventilation so that the bees can regulate the colony's temperature. In addition, modern hives provide the beekeeper with the ability to inspect, manipulate, and manage the colony. So exactly what kinds of conditions cater to these necessities?

The Langstroth hive is the most widely used hive in the United States, and it's gaining popularity worldwide. In the following sections, I use it to illustrate the basic components of a hive and their function; check them out in Figure 2-2. (You can find out how to build a Langstroth hive in Chapter 10.)

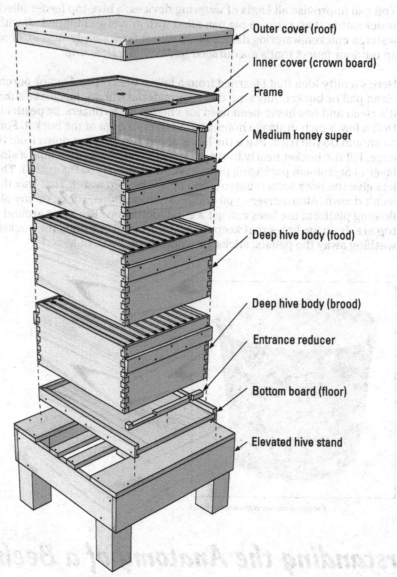

Outer cover (roof)

Inner cover (crown board)

Frame

Medium honey super

Deep hive body (food)

Deep hive body (brood)

Entrance reducer

Bottom board (floor)

Elevated hive stand

Figure 2-2:
The basic compo-
nents of a
Langstroth
hive.

Illustration by Felix Freudzon, Freudzon Design

Elevated hive stand

A hive stand isn't necessary, but you may find one useful because it elevates
the hive off the wet ground, which improves air circulation and results in less
bending over when you're inspecting your hives. In addition, grass growing
in front of the hive's entrance can slow the bees' ability to get in and out.
The stand alleviates that problem by raising the hive above the grass. See
Chapter 13 for instructions on building an elevated hive stand that you can
use with a Langstroth, British National, or nuc hive.

Bottom board

A *bottom board* is the floor of the beehive. It consists of several rails that serve as a frame around a solid piece of wood, and it protects the colony from damp ground. These days, more and more beekeepers are using what's called a *screened* bottom board in place of the standard bottom board. This improves ventilation and is helpful when controlling and monitoring the colony's population of varroa mites. Flip to Chapter 14 for instructions on how to build a screened bottom board.

Entrance reducer

An *entrance reducer* is a cleat that limits bee access to the hive and controls ventilation and temperature during cooler months. You don't nail the entrance reducer into place but rather place it loosely at the hive's entrance so that you can introduce it or remove it as needed. The small notch reduces the entrance of the hive to the width of a finger. The large notch opens the entrance to about four finger widths. Removing the entrance reducer completely opens the entrance to the max.

If the hive design you choose doesn't use an entrance reducer, you can use clumps of grass to close off some of the entrance.

Deep hive bodies

The deep hive bodies are essentially boxes that contain frames of comb. For a Langstroth hive, you typically build two deep hive bodies to stack on top of each other, like a two-story condo. The bees use the *lower deep* as the nursery or *brood chamber*, to raise thousands of baby bees. They use the *upper deep* as the pantry or *food chamber*, where they store most of the honey and pollen for their use.

If you live in an area where frigid winters just don't happen (temperatures don't go below freezing), you may not need more than one deep hive body for your colony (one deep for both the brood and their food). In such situations, you want to monitor the colony's food stores and feed the bees if their supplies run low.

Honey super

Beekeepers use *honey supers* to collect surplus honey. That's *your* honey — the honey that you can harvest from your bees. The honey that's in the deep hive body must be left for the bees. Supers are identical in design to the deep hive bodies (see the preceding section), and you build and assemble them in a similar manner. But the depth of the supers is more shallow.

Honey supers typically come in two popular sizes: shallow (which usually measure 5¾ inches high) and medium (which usually measure 6⅝ inches high). Medium supers are sometimes referred to as *Illinois* supers because they were originally developed by Dadant & Sons, Inc., which is located in Illinois. I prefer medium supers to shallow supers and use mediums exclusively. Why? The mediums hold more honey and yet are still light enough that I can handle them fairly easily when packed with golden goodness (medium supers weigh in at around 50 to 55 pounds when packed full). However, many of my beekeeping friends use shallow supers because they're just that much lighter when filled with honey (they weigh around 35 to 40 pounds when packed full). For this reason, the plans in Chapter 10 for the Langstroth hive provide cut sheets for building either medium or shallow honey supers. The choice is yours.

You can use medium-size equipment for your *entire* Langstroth hive (no deeps). Three medium-depth hive bodies is about equivalent to two deep hive bodies. Standardizing on one size means that all your equipment is 100 percent interchangeable. The lighter weight of each medium hive body makes lifting much, much easier than manipulating deep hive equipment (in comparison, deep hive bodies can weigh up to 100 pounds when full).

As the bees collect more honey, you can add more honey supers to the hive, stacking them on top of each other like so many stories to a skyscraper. For your first season, build one honey super. In your second year, you'll likely need to build two or three or more supers. Honey bonanza!

Frames

Four of the hives in this book use removable frames (nuc, observation, British National, and Langstroth). The bees build their honeycomb onto the frames. Because the frames are removable from the hive, you can easily inspect, manipulate, and manage the colony. For the nuc, observation, and Langstroth hives, the wooden frames contain a single sheet of beeswax foundation. Frames typically come in three basic sizes: deep, shallow, and medium, corresponding to deep hive bodies and shallow or medium honey supers.

You can certainly purchase frames from a beekeeping supply vendor. Or you can find out how to build your own Langstroth-style frames from scratch in Chapter 17. You use them with the nuc hive in Chapter 6, the four-frame observation hive in Chapter 7, and the Langstroth hive in Chapter 10. You use simpler top bars with the Kenya top bar hive and the Warré hive; details on building those are in Chapters 5 and 8. The British National hive uses top bar hybrid frames; get the scoop on how to build them in Chapter 9. You can see a Langstroth-style frame and a top bar in Figure 2-3.

Inner cover

The inner cover of the hive resembles a shallow tray (with a ventilation hole in the center). I also like to cut a notch in one of the short lengths of the frame. This is an extra ventilation source, positioned to the front of the hive. You place the inner cover on the hive with the tray side facing up (see Figure 2-4).

Figure 2-3:
Langstroth-
style frames
(left) and top
bar frames
(right).

Photographs courtesy of Howland Blackiston and Jim Fowler

Figure 2-4:
For the
correct
placement
of an inner
cover, the
tray side
faces up.

Photograph courtesy of Howland Blackiston

Alternatively, screened inner covers have been gaining popularity in recent years. They provide the colony with terrific ventilation. Skip to Chapter 12 for plans on building a double screened inner cover.

You do *not* use the inner cover at the same time you have a hive-top feeder on the hive. You use the hive-top feeder in place of the inner cover. Chapter 15 has plans for building a hive-top feeder.

Outer cover

The outer cover protects the bees from the elements. Like the roof on your house, you can ensure that it's waterproof and also extend the life of the wood by covering the top with a weatherproof material (aluminum flashing, asphalt tiles, cedar shingles, and so on). See Chapter 19 for some alternative roof ideas.

Looking at Locale

Before you build a terrific new beehive, you need to know where you're going to put it. You should become acquainted with legal, neighborly, and venue considerations, as I describe in the following sections.

Following regional laws and requirements

Is it legal to keep bees? In most places, the answer is yes. But some areas have laws or ordinances restricting or even prohibiting beekeeping. For the most part, such restrictions are limited to highly populated, urban areas. But even that is changing. In 2010, New York City lifted a long-standing ban on beekeeping. Now beehives are popping up all over the city! The chef at New York's Waldorf Astoria even keeps a few hives on the roof of the hotel.

Some communities may limit the number of hives you can keep, and some require you to register your bees with town/city hall. To find out the legality of keeping bees in your area, contact your town/city hall, state bee inspector, state agricultural experiment station, or a local bee club; a quick online search should yield the contact information you need.

Bee Culture magazine maintains a terrific online listing of regional clubs in the United States and Canada. Visit www.beeculture.com and follow the link to "Find a Beekeeper Near You." These clubs and associations can be very helpful in identifying whatever local laws and legislations apply in your region.

If you live in an apartment, speak to your landlord about roof rights. See whether you can get access to your building's rooftop and obtain permission to place a hive or two on the roof. Rooftop hives are wonderful because they're out of sight to most people, which reduces neighborhood fear and lessens the chance of vandalism.

Bee-ing sweet to your neighbors

Whether you live in the country or the city, for many among the general public, ignorance of honeybees is complete. Having been stung by hornets and yellow jackets, they assume having any kind of bee nearby spells trouble. Not true. It's up to you to take steps to educate them and alleviate their fears.

Here are some things you can do to put your neighbors at ease:

✔ Restrict your bee yard to one or two hives, particularly if you're in an urban setting. Having only a couple of hives is far less intimidating to the uneducated than having a whole phalanx of hives.

✔ Locate your hive in such a way that it doesn't point at your neighbor's driveway, your house entrance, or any other pedestrian traffic-way. Bees fly up, up, and away as they leave the hive. When they get 15 feet from the hive, they're way above head level.

✔ Don't flaunt your hives. Put them in an inconspicuous area.

✔ Paint or stain your hives to blend into the environment. Painting them flame orange is only tempting fate.

✔ Provide a nearby source of water for your bees. That keeps them from collecting water from your neighbor's pool or birdbath (see the earlier section "A nearby source of water" for more info).

✔ Invite folks to stop by and watch you inspect your hive. They'll see first-hand how gentle bees are, and your own enthusiasm will be contagious.

✔ Let your neighbors know that bees fly in about a 3-mile radius of home plate (that's roughly 6,000 acres). So they'll mostly visit a huge area that isn't anywhere near your neighbor's property.

✔ Give gifts of honey to all your immediate neighbors and/or your land-lord (see Figure 2-5 for an example). This gesture goes a long way in the public relations department.

Figure 2-5: This gift basket of honeybee products goes to each of my immediate neighbors. That's sure to help keep the peace.

Photograph courtesy of Howland Blackiston

Picking the perfect location

You can keep bees just about anywhere: in the countryside, in the city, in a corner of the garden, by the back door, in a field, on the terrace, or even on an urban rooftop. You don't need a great deal of space or flowers on your property; bees happily travel for miles to forage for what they need. These girls are amazingly adaptable, but you'll get optimum results and a more rewarding honey harvest if you follow some basic guidelines, as you discover in this section. (I provide some special tips for those of you in urban settings, too.)

Location fundamentals, no matter where you live

The ideal hive location has easy access (so you can tend to your hives), good drainage (so the bees don't get wet), a nearby water source for the bees, dappled sunlight, and minimal wind (see Figure 2-6). Keep in mind that fulfilling *all* these criteria may not always be possible. No worries — the bees will forgive you. Do the best you can by following these basic guidelines:

✔ Face your hive to the southeast. That way your bees get an early morning wake-up call and start foraging early.

✔ Position your hive so that it's easily accessible come honey harvest time. You don't want to be hauling hundreds of pounds of honey up a hill or down a fire escape on a hot August day.

✔ Provide a windbreak at the rear of the hive. I've planted a few hemlocks behind my hives. Or you can erect a fence made from posts and burlap or even use bales of hay to block harsh winter winds that can stress the colony (assuming you live in a climate with icy-cold winters).

✔ Put the hive in dappled sunlight. Full, direct sun all day long causes the hives to get very hot in the summer. The bees will spend valuable time trying to regulate the hive's temperature (rather than making honey). You also want to avoid deep, dark shade because it can make the hive damp and the colony listless.

✔ Make sure the hive has good ventilation. Avoid placing it in a gully where the air is still and damp. Also, avoid putting it at the peak of a hill, should you live in a region where the bees will be subjected to winter's fury.

✔ Place the hive absolutely level from side to side, with the front of the hive just slightly lower than the rear (a difference of an inch or less is fine), so that any rainwater drains out of the hive (and not into it).

✔ Locate your hive on firm, dry land. Don't let it sink into the quagmire.

In a country setting, you can place mulch around the hive to prevent grass and weeds from blocking its entrances.

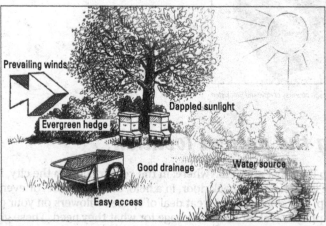

Figure 2-6:
The perfect suburban setting for your hives has easy access, good drainage, a nearby water source, dappled sunlight, and minimal wind.

Prevailing winds
Dappled sunlight
Evergreen hedge
Good drainage
Water source
Easy access

Illustration courtesy of Howland Blackiston

Special urban considerations

Just about all the considerations listed in the preceding section apply to urban situations. Here are a few more details for all you city beekeepers out there, according to New York City beekeeper Andrew Coté.

✔ **Decide upfront where to put your hives.** Placement of urban hives is often tricky and a stumbling block for many metropolitan beekeepers. Don't be one of those beekeepers who takes a course, builds a hive, gets a package of bees, and then realizes there's no suitable place to put the bees! Do your homework upfront.

✔ **Strike a deal with a community garden.** These are usually run by small neighborhood groups who are sympathetic to honeybees, welcome their pollination, and are likely eager to offer a home for your hives. A search on the Internet will quickly find community gardens in your neighborhood.

Don't be so grateful for a spot in the community garden that you impulsively offer half of your honey harvest! Your "rent" should be bartered in exchange for the considerable pollination services you bring to the garden. The honey should be all yours.

✔ **Stay safe on the roof.** Though a roof is a great location for urban beehives (check out Figure 2-7 for an example), you need to be aware of some safety issues:

- Avoid a roof if you have to go up a fire escape, climb a tall ladder, or use a rooftop hatch. In all these situations, attempting to remove full and heavy honey supers from the roof area would be difficult and dangerous.

- Don't place your hive too close to the edge of the roof. If you end up with a bee up your pant leg and you lose your balance, no amount of arm-flapping will help you fly safely to the ground.

- Secure all the parts of the hive using crank straps. Strong gusts of wind can send hive parts flying wildly off the roof to pedestrians below. And there's more wind up on those roofs than you may realize.

Never place a beehive on a fire escape. *Never.* It's illegal and it's dangerous.

Figure 2-7: A rooftop is a great location for urban hives. Note the rocks to keep covers from blowing off, bales of hay as a windbreak, and a water source placed in front of the hives.

Photograph courtesy of Andrew Coté, New York City Beekeepers Association

Making a Beeline to the Hive That's Best for Your Needs

So with all the different hives in this book to choose from, how do you decide which one to build? Maybe you just like the look of one hive over another. A better way to decide is to determine the primary reasons you're beekeeping and select the hive best suited for those reasons. You also need to consider your level of woodworking experience. The following sections can help you make these decisions.

For a better understanding of the advantages and disadvantages associated with the various hives in this book, see the introduction and the "Vital Stats" section in each of the chapters in Part II.

A hive for learning and teaching

Suppose you're really interested in bees but don't want to deal with all that outdoors stuff every week. Your primary interest is to learn more about bees — to study their behavior and observe the fascinating things that take place inside a hive. Kind of like having an aquarium and watching the fish do their thing. You feed them a little and occasionally clean the glass on the tank, but that's about it.

Or maybe you want to make presentations at schools, nature centers, and farmers markets, and you need a hive that's portable and can safely display live bees.

The choice for you is the four-frame observation hive (see Chapter 7). It's a terrific portable hive for show and tell and for, well, observing.

Hives for pollinating your garden

Suppose your primary reason for having bees is to improve pollination in your garden. You don't care about harvesting honey. You don't care about a show and tell hive. You just want larger and more abundant flowers, vegetables, and fruits. Bees can make that happen.

The good news is that any of the hives in Part II helps pollinate a garden. But they accomplish this to varying degrees. The larger the hive, the larger the bee population for pollination but the more work for you. So if your intent is to max-out pollination, consider the Kenya, Warré, British National, or Langstroth hives (Chapters 5, 8, 9, and 10, respectively). If you don't want all the work associated with larger hives, a nuc hive tucked into the corner of your garden or fruit orchard will likely do a decent job of pollinating during the growing season (see Chapter 6).

A hive for harvesting honey

Ah, honey. Maybe you just want tons of that liquid gold!

The Langstroth is likely the hive for you. Though *all* the larger hives produce honey you can harvest, the Langstroth is the granddaddy of honey producers (see Chapter 10).

Hives to match your building skills

Maybe you're fairly new at carpentry, so you think you should start with a few easy builds and work your way up.

Though experienced woodworkers can jump right in and tackle any of the hives or equipment listed in this book, new woodworkers may want to get their feet wet by starting with some easy builds. The Kenya top bar hive, the nuc hive, and the four-frame observation hive (with purchased frames) are easy builds (see Chapters 5, 6, and 7).

Hives for selling

Perhaps you're an experienced woodworker and would like to make hives and bee equipment to sell to other beekeepers. So, which of this book's designs would be the most marketable?

The Langstroth hive in Chapter 10 is the most widely offered beehive in the world. The commercial beekeeping suppliers make these hives by the thousands, and competing with their discounted pricing would be difficult. The most marketable hives and equipment are those not typically offered by the major bee suppliers. Consider making and selling the hives that are gaining popularity but aren't widely available, such as the Kenya hive (Chapter 5) and the Warré hive (Chapter 8). These can be very marketable, particularly if you jazz 'em up with some special design details (turn to Chapter 19 for ten ways you can trick out the hives you intend to sell).

A handy table to help you decide the hive to build

So which hive should you build? Table 2-1 helps you decide on the hives that best meet your beekeeping needs and your woodworking skills.

Table 2-1 Deciding Which Hive(s) Best Meets Your Needs and Skills

Hive Type	Show & Tell	Pollination	Honey Extraction	Building Difficulty	Marketability	Notes
Kenya top bar hive (Chapter 5)	No	Good	Good	Easy	Good	A great hive for those seeking the most natural environment for their bees. Very cost-effective to make. There's a growing popularity for this hive, which makes it very marketable for sale.
Nuc hive (Chapter 6)	No	Moderate	N/A	Easy (if you purchase frames); moderate (if you build your own frames)	Poor	This is a very small hive. It's ideal for starting a new colony, raising queens, or providing some pollination in a garden.
Four-frame observation hive (Chapter 7)	Yes	Poor	N/A	Easy (if you purchase frames); moderate (if you build your own frames)	Fair	Perfect for portable show-and-tell and for observing behavior at home. Not intended for honey production. Bees must be fed throughout the year. Bees must be permitted free flight when not being used for educational display.
Warré hive (Chapter 8)	No	Good	Good	Moderate	Good	The peaked roof and some details at the entrance add a modicum of complexity to this otherwise easy build. Has the smallest footprint of all the exterior hives in this book, which makes it ideal for someone with very limited space. There's a growing popularity for this hive, which makes it very marketable for sale.
British National hive (Chapter 9)	No	Good	Very good	Moderate	Poor	My design uses a top bar configuration to simplify the build and because foundation for this size frame isn't readily available outside of the United Kingdom.
Langstroth hive (Chapter 10)	No	Good	Very good	Moderate (if you purchase frames); difficult (if you build your own frames)	Poor	Building frames and making finger joints can be a challenge to the neophyte woodworker.

Chapter 3

Gathering Basic Tools and Materials

· ·

In This Chapter

▶ Playing it safe

▶ Selecting the tools you'll need

▶ Picking the right lumber

▶ Rifling through other materials

▶ Focusing on the finishing touches

▶ Estimating materials and costs

· ·

You don't need to be a master carpenter to build the hives and equipment featured in this book. Though some of the designs may be a tad trickier than others, none is anything as complicated as, say, building a house, or even a chicken coop. As long as you're comfortable working with some basic power tools, building your own beekeeping equipment is a fun hobby that will give you a great sense of accomplishment and provide your bees with some really terrific digs.

The plans in this book use basic tools and materials. You need not have a professional workshop stocked with thousands of dollars of fancy woodworking equipment. In fact, I've intentionally kept the number of different hand and power tools to a bare minimum. This chapter gives you some safety tips, runs down the tools and materials you need, and gives you ideas for putting the finishing touches on your handiwork.

Bee-ing Safe Before You Begin

Before you pick up a hammer to drive a single nail, spending a few moments to review some safety basics is essential. Any sharp tool has the potential to create the kind of danger in your workshop that can be far worse than a bee sting in the garden. Using a little common sense and taking a few simple precautions should be the first step in your beehive-building adventure.

Protecting yourself with safety gear

The most common injuries in woodworking don't come from dramatic accidents involving steel and flesh but rather from subtle things that happen over time to damage breathing, ears, and eyes. The following safety gear is essential to any woodworking workshop. Always wear this safety gear, no matter how silly you think it makes you look.

✔ **Breathing protection:** Cutting, drilling, and sanding wood can generate lots of airborne sawdust and wood dust and do some long-term damage to your lungs. Some woods are particularly troublesome, such as cedar. Here are a couple of options to protect you (see Figure 3-1):

- **Dust masks:** These are inexpensive and disposable and can be found at any hardware store or home center. They look like a surgeon's mask and work well to keep out the dust (but not the fumes associated with paint and varnish).

- **Respirators:** These aren't real fashionable looking, but they're excellent at keeping out dust and also the fumes associated with paints, polyurethanes, and varnishes. Just be certain that the model you choose is rated to protect you against the materials you're using.

✔ **Ear protection:** Hammering nails and operating power tools can get noisy and seriously damage your hearing over time. A router cutting into wood can produce over 110 decibels — that's equal to a deafening rock concert. You have a couple of simple options to protect your hearing (see Figure 3-2):

- **Earplugs:** These soft plugs insert right into your ear canal and are typically made of foam or plastic. They're inexpensive but can sometimes be difficult to fit and can pop out or loosen, resulting in diminished effectiveness.

- **Earmuffs:** These completely cover your ears (like large stereo headphones) and dramatically shut out all ambient noise. They're far more effective than earplugs but can be tiresome to wear over long periods.

One great advantage of earmuffs is that you get them with a built-in radio or a wireless connection to your portable music player. So you can listen to Nikolai Rimsky-Korsakov's "Flight of the Bumblebee" while you build your hives. Just be sure not to turn the volume up too loud or you'll defeat the purpose of wearing them.

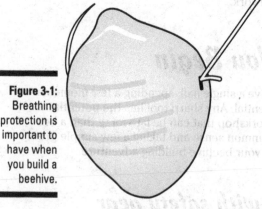
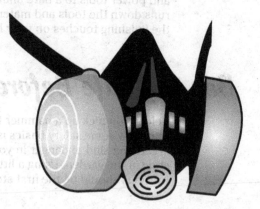

Figure 3-1: Breathing protection is important to have when you build a beehive.

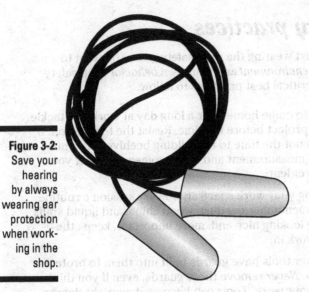
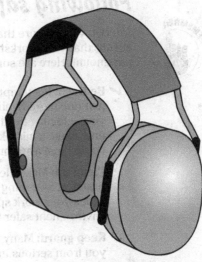

Figure 3-2:
Save your hearing by always wearing ear protection when working in the shop.

Illustration by Wiley, Composition Services Graphics

✔ **Eye protection:** Seriously, this is one bit of safety gear you should always make use of. By their very nature, woodworking and power tools mean all kinds of dust and wood flecks flying all about. Safety glasses or goggles are a must (check out a pair in Figure 3-3), and some rather stylish ones are available today. So don't be vain — put these protective glasses on whenever you're working in your shop. Most are designed to fit right over prescription eye glasses.

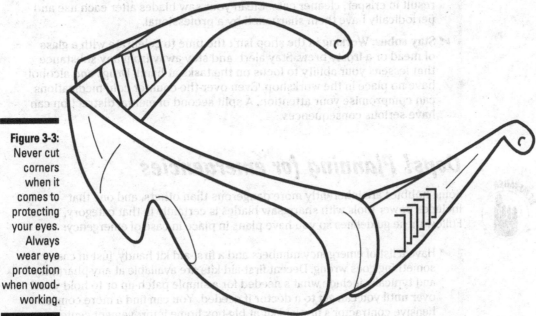

Figure 3-3:
Never cut corners when it comes to protecting your eyes. Always wear eye protection when woodworking.

Illustration by Wiley, Composition Services Graphics

Following safety practices

Safety involves more than just wearing the right safety gear. You want to ensure that your workshop *environment* and your own *behavior* keep safety paramount. Here are some critical best practices to follow:

- **Be fresh:** It's tempting to come home after a long day at work and tackle just one more beehive project before bedtime. Resist the temptation. If you're feeling tired, it's not the time to be building beehives. You may only make a mistake in measurement and waste some wood, but you could have a serious accident.

- **Clean up:** Make cleaning your work space after every session a routine practice. Sweeping or vacuuming sawdust, wood chips, and liquid spills keeps your work space looking nice and, more important, keeps the environment safer to work in.

- **Keep guard:** Many power tools have guards built into them to protect you from serious injury. *Never* remove these guards, even if you think they get in the way of your work. Tools can become downright dangerous without them.

- **Light up:** Provide yourself with as much light as possible in your shop. Invest in some bright shop lights to fully illuminate your work area as you operate your power tools.

- **Slow down:** Building beehives isn't a race. Take your time. Haste makes waste, my mom always said. If you're in a hurry, you'll certainly make mistakes or, worse, cause injuries. Going slowly results in easier and more accurate cuts and a better-looking project.

- **Stay sharp:** Clean, sharp tools are not only easier to work with but also result in crisper, cleaner cuts. Clean your saw blades after each use and periodically have them sharpened by a professional.

- **Stay sober:** Working in the shop isn't the time to celebrate with a glass of mead or a frosty brew. Stay alert, and stay away from any substance that lessens your ability to focus on the tasks at hand. Drugs and alcohol have no place in the workshop. Even over-the-counter cold medications can compromise your attention. A split second of mental distraction can have serious consequences.

Oops! Planning for emergencies

Some hobbies are inherently more dangerous than others, and one that includes power tools with sharp saw blades is certainly in that category. Follow these guidelines so you have plans in place in case of emergency:

- Have a list of emergency numbers and a first-aid kit handy, just in case something goes wrong. Decent first-aid kits are available at any pharmacy and typically include what's needed for a simple patch-up or to hold you over until you can get to a doctor if needed. You can find a more comprehensive contractor's first-aid kit at big-box home improvement centers.

✔ Be sure to let someone in the house know you're working in the shop or make sure a neighbor is around to help you if necessary.

✔ Have a fire extinguisher readily available where you do your woodworking.

Talking about Tools

Chances are that some of the tools you need to build a beehive are already somewhere in your garage, basement, or that catch-all kitchen drawer (along with those nuts and bolts of mysterious origin). I describe these essentials, along with a few other goodies, in the following sections.

A few essential hand tools

Here are some basic hand tools you need for nearly all the build plans in this book:

✔ **Brad driver:** Sometimes called a *brad pusher,* this spring-loaded tool (see Figure 3-4) allows you to push a small brad into wood. No hammering and no smashed thumbs! Just slip a finish brad nail into the spring loaded tube and press the tube against the wooden part you're attaching. The little brad slides right into place. This is a very handy tool when assembling frames and installing foundation. It makes short work of attaching the wedge bar that holds the foundation in place.

✔ **Carpenter's hammer:** You can find a type of hammer for just about every application, but what you want for these building projects is a *carpenter's hammer* with a 16 to 20 ounce head. Chances are the hammer you already have at home falls into this category. If you'll be making lots of beehives and equipment, be sure to invest in a quality hammer (with good balance and a good grip). It costs a bit more than a standard hammer, but there's a big difference in ease of use.

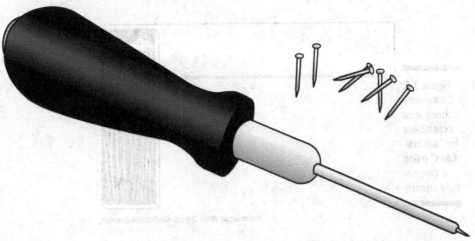

Figure 3-4:
A brad driver is an excellent tool when it comes to assembling frames and foundation.

Illustration by Wiley, Composition Services Graphics

✔ **Carpenter's pencil:** I know, I know. This may not be *essential.* Any pencil can mark measurements on your lumber. But a *carpenter's pencil* is over-sized (making it easier to handle) and flat, not round, meaning it won't roll off the worktable, resulting in time on your hands and knees looking for it. For under half a dollar, it's worth the investment (see Figure 3-5).

✔ **Carpenter's square:** Sometimes called a *try square,* this is a must-have tool to ensure that your assemblies are true and square (see Figure 3-6). If they're not, you'll have some seriously wobbly equipment. You can get various length blades on a carpenter's square, but the 8 to 12 inch length is most useful for these projects. Check out Chapter 4 for instructions on how to use a square.

✔ **Paintbrushes:** You'll probably want to protect your woodenware from the elements by applying a good quality outdoor paint, an exterior polyurethane, or marine varnish. Therefore, you need a variety of paintbrushes. I like to use inexpensive disposable brushes (there's no cleanup!). I typically get 2 inch and 4 inch wide brushes. True, cheap brushes can sometimes leave a few loose bristles in the finish, but hey, you're making beehives, not museum furniture.

Figure 3-5:
The fat design of a carpenter's pencil makes it easy to handle, and being flat keeps it from rolling off your worktable.

Illustration by Wiley, Composition Services Graphics

Figure 3-6:
Continually check your assemblies for "square-ness" using a carpenter's square.

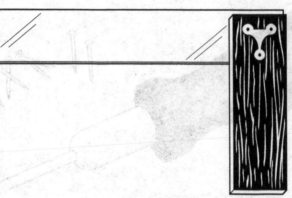

Illustration by Wiley, Composition Services Graphics

✔ **Power drill:** You can use a power drill with a cord, but the cordless, battery-powered version is my favorite for portability and ease of use. The 18- or 20-volt models hold a good charge and have lots of *torque* (twisting power). Most cordless drills let you select the amount of torque (see Figure 3-7). Dial up more torque when you need more driving power to set a screw; dial it down if your screws are going too deep or you're stripping the head of the screw.

You use a power drill for two tasks:

- **To create starter holes:** These holes are for the fasteners (screws and nails) you use. Pre-drilling makes inserting fasteners easier and helps prevent the wood from splitting. You need a ⁵⁄₆₄-inch drill bit for most of the pre-drills. There are other types of holes needed in the book. Some of the hive designs call for a 1-inch drill bit, and the four-frame observation hive (see Chapter 7) calls for 1½-inch and 3-inch drill bits. Also, you need a ⅛ inch drill bit for making frames (Chapters 9 and 17).

- **To attach screws:** I typically use two types of screws throughout this book — deck screws and lath screws. To keep it simple, I've specified all the screws used in the book with #2 size Phillips drive. So you need a #2 size Phillips head bit for your power drill.

✔ **Staple gun:** Some of the projects involve attaching hardware cloth and other materials to hive and equipment parts. A hand-held, heavy-duty staple gun is the perfect match for the job. Add a supply of ⅜-inch staples and you're good to go.

✔ **Tape measure:** A standard, retractable, metal tape measure is just fine for the projects in this book. Because the measurements in this book are in inches (versus metric), make certain your tape measure is calibrated in inches.

✔ **Tin snips:** Some of the beehive designs use galvanized metal hardware cloth for ventilation and to keep the bees contained in an area. In addition, some of the roof assemblies call for aluminum flashing. Chances are your kitchen scissors or that pair of pliers in your toolbox won't make much headway when it comes to cutting these metals. You need something beefier. A good pair of tin snips (see Figure 3-8) will slice through this stuff with ease.

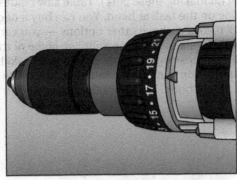

Figure 3-7: Most cordless drills let you select the amount of torque.

Illustration by Wiley, Composition Services Graphics

Figure 3-8:
Tin snips cut through rugged hardware cloth and aluminum flashing like a hot knife through butter.

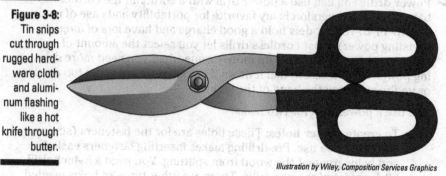

Illustration by Wiley, Composition Services Graphics

The right saws and blades

Forget about old-fashioned handsaws. You won't likely use them on any of the projects in this book. They're slow going, tiresome to use, and simply not precise enough for the task. Accuracy is important when assembling the boxes that become your hives. And for accuracy, nothing can compete with power saws. For the projects in this book, you need only two kinds of power saws and a small collection of blades.

✔ **Circular saw:** Sometimes called a *Skilsaw* (named for the company that invented it), this hand-held power tool (see Figure 3-9) is lightweight and easy to control. You can change out blades to accommodate whatever material you're cutting through. Technically speaking, a skilled woodworker could use a circular saw for most of the builds and joinery cuts in this book. But I much prefer using a table saw for almost all the cuts in the book (see the next bullet). I reserve my circular saw not for making precise cuts but for quickly trimming large sheets or long boards of lumber down to more manageable sizes. In all likelihood, you can get by with only a *combination blade* for your circular saw.

✔ **Table saw:** If you don't have a table saw (see Figure 3-10), I urge you to invest in one for the projects in this book. When building beehives, you use this tool more than any other. For making straight and precise cuts, a table saw is much easier to work with than a circular saw. And it's irreplaceable when it comes to making rabbet cuts and dado joints (see Chapter 4 for information on making these cuts). Table saw blades are interchangeable depending on the task at hand. You can buy a decent table saw for less than $300. You do have other options — you can use chisels and hand saws and circular saws to accomplish some of the same objectives. But some of these options require advanced skills in woodworking. Keep it simple. You can use the table saw for every plan, cut, and joinery covered in this book. You'll wonder how you ever did without one.

✓ **Table saw blades:** You need a few different types of blades for your
table saw, each having a specific purpose. Invest in good-quality carbide
blades; they hold their edge longer and are safer and easier to work with
(a sharper blade means less effort on your part).

You'll use three blade types for your table saw for the projects in this
book (see Figure 3-11):

- A combination saw blade (for cutting with or against the grain)

- A plywood saw blade (for cutting plywood)

- A stacked set dado saw blade, adjustable up to ¾ inch width (for
  making rabbet cuts and finger joints)

Figure 3-9:
A circular
saw is
handy for
trimming
larger
pieces of
lumber
down to
more man-
ageable
sizes.

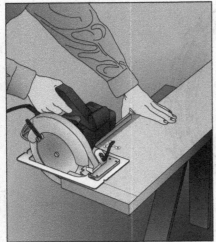

Illustration by Wiley, Composition Services Graphics

Figure 3-10:
Having a
table saw
is a huge
benefit when
it comes to
building
beehives.

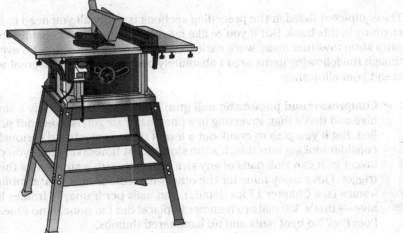

Illustration by Wiley, Composition Services Graphics

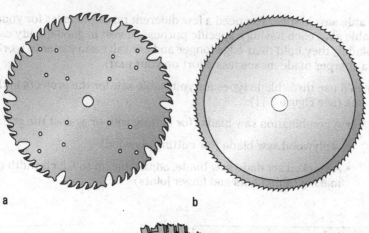

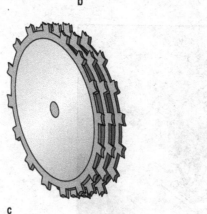

a   b

Figure 3-11:
You use
three blade
types for
your table
saw: (a)
combination
saw blade,
(b) plywood
saw blade,
and (c)
stacked set
dado saw
blade.

c

Illustration by Wiley, Composition Services Graphics

Gadgets that are nice to have but not essential

The equipment listed in the preceding sections is really all you need to build anything in this book. But if you're like me, it's mighty fun to consider those extra shop toys that make work easier, faster, or more enjoyable. So even though the following items aren't absolutely necessary, they're a great way to spend your allowance.

✓ **Compressor and pneumatic nail gun:** If you plan to build only a single hive and that's that, investing in a pneumatic air gun is likely not justi-fied. But if you plan to crank out a few of these projects, I'd seriously consider looking into this. It's the single most timesaving tool you can invest in. It can sink nails of any size or length with a squeeze of the trigger. I love using mine for the otherwise tedious task of assembling frames (see Chapter 17 for details). Ten nails per frame, 30 frames in a hive — that's 300 nails to hammer in place! But I'm done in no time. Pop! Pop! Pop! No bent nails and no hammered thumbs.

As helpful as nail guns can be, the word *gun* should be enough to raise a safety flag. Like any power tool, nail guns present an opportunity for danger. Here are some safety hints:

- Before using, carefully read the owner's manual.
- Never tamper with or remove the built-in safety devices.
- When loading the gun with nails, disconnect the gun from its power source.
- Always wear soundproof ear protectors and safety glasses.
- Keep your finger off the trigger if you're not firing the tool.
- Never point the gun in a direction you don't intend for a nail to go.

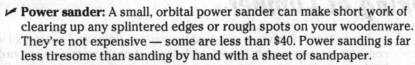

Complete nail gun starter kits run about $250. If you're not quite ready to invest in this luxury, see if you can borrow one from a friend, or consider renting one.

✔ **Power sander:** A small, orbital power sander can make short work of clearing up any splintered edges or rough spots on your woodenware. They're not expensive — some are less than $40. Power sanding is far less tiresome than sanding by hand with a sheet of sandpaper.

✔ **Shop vacuum:** You're going to generate some serious sawdust from your power tools. In a perfect world, your shop would include a dust collector system. But in the meantime, a *shop vac* (wet/dry vacuum) will help you quickly clean up the mess during and after your woodworking. You can use the vac to directly connect to hoses in your power equipment and suck up the dust *before* it becomes airborne. A good wet/dry shop vacuum with a built-in fine particle filter starts at about $130.

In the following list, I'm talking strictly fun stuff. Though not a must, how cool are these gadgets?

✔ **Custom branding iron:** With this gadget you can mark your products with your name, initials, or a logo. A number of companies on the Internet make custom branding irons. Some are electric, and some are propane-powered. Search on the web for "custom branding irons." Using one is a fun and proprietary way to brand your handiwork as your own.

✔ **Dust collector:** A shop vacuum with a fine particle filter is a good start to keeping your shop clean and the air safer, but if you're bashing out the beehives, you'll want to have a proper dust collector system in your shop. It's essentially a whacking big vacuum that connects by hoses to all your shop equipment. The unit captures the dust and wood chips that your shop tools generate. Dust collectors can be a bit pricey ($150 to more than $600), but they do an excellent job. Your lungs will thank you.

✔ **Router table:** In the simplest terms, a router table consists of a sturdy flat table surface with a router attached to the bottom. The router itself is a motor with interchangeable bits. The router is mounted with its base attached to the underside of the table and the cutting bit sticking up through a hole in the table's surface. The table holds the router securely, leaving both hands free to guide the lumber being cut. By changing out the bits, you can easily perform all the tasks necessary to cut dadoes, rabbets, finger joints, grooves, and a host of other nifty woodworking tasks. Though you can certainly accomplish these same cuts using a table saw, many woodworkers prefer using a router and table. If you have the space and budget, you won't ever regret having one in the shop. Decent tables with a router motor start at around $100.

✔ **Self-marking tape measure:** A self-marking tape measure is a tape measure with a marking cartridge that eliminates the need for a pencil or an additional marking tool. At the point where the tape exits the roll, there's a pre-inked marker on the underside. Just set the tape measure down on the lumber you're measuring, and it accurately makes the mark. Awesome. I see them online for less than $20.

Looking at Lumber

In this section I review a number of the various types and species of wood to consider for your beehive projects. I also sort out how to size and order your lumber.

Choosing lumber

Beehives, including all the ones in this book, are typically made from wood. You have hundreds of kinds of wood to choose from. But from a practical and financial viewpoint, you should limit your "discovery" to those woods that are readily available from most lumberyards and home improvement centers. Some woods are very cost-effective (they're cheap), and some are very exotic (they're expensive). In the following sections, I note a few options for your lumber.

The best woods for beehives

A handful of woods are used most frequently for making beehives. Sometimes the choice is regional (selecting a wood that's readily available in your area), sometimes the choice is financial (selecting a wood because it's the lowest price), and sometimes the choice is based on durability (selecting a wood because it stands up to the elements better than other woods). What follows are the woods most frequently used by the commercial manufacturers of beekeeping equipment.

✔ **Pine:** Hands down, this is the most widely used choice. It's readily available everywhere, it's among the least expensive lumber to purchase, and it's easy to work with. For this reason, nearly all the builds in this book recommend pine boards as the lumber of choice. Note that there are different grades of pine. Two are of interest here:

• **Knotty (sometimes referred to as *standard* pine):** The knotty grade of pine is the least expensive and is perfectly sound for building, but it does contain knots and some other cosmetic imperfections. If you can live with a more rustic-looking hive, by

all means go for the standard grade and save a few bucks. You
may need to order a little more material than stated in the plans,
as the knots sometimes fall right where you plan a critical piece of
joinery. So having a little extra material allows you to choose an
alternate piece of wood without the offending knot.

- **Clear (sometimes referred to as *select* or *premium* pine):** The clear
grade is pricier. As the name suggests, this grade is clear of knots
and other blemishes. The grain of the wood also tends to be tighter
and straighter. It makes for a very nice looking, defect-free hive.

Pine isn't the most durable wood to weather the elements, so you must
protect pine with some coats of outdoor-quality paint, exterior polyurethane, or marine varnish. (See the later section "Protecting Your
Beehives with Paint and More" for details.)

✔ **Cypress:** The cypress tree produces a sap-type oil that preserves the
wood and naturally repels insects and mold. So cypress is a terrific
wood for making beehives and beekeeping equipment. But with most
of the wood coming from southern states in the United States, it's not
readily available all across the country. But if you can get your hands on
some cypress, you won't be disappointed in the results. It's a beautiful
and naturally durable wood for building beehives.

✔ **Cedar:** Cedar is a beautiful wood, and it smells divine. The natural
oils make it less prone to warping, less susceptible to bug infiltration,
and less likely to rot than other woods. Though you can paint it, you
certainly don't have to because of its naturally durable qualities. Left
untreated it will weather to a lovely, light gray patina. Frankly, were it
not for the fact that it's more expensive than pine, I'd use it for any hive
listed in this book. (For the British National hive in Chapter 9, I specify
cedar because in Europe, cedar is traditionally used for this hive.) Many
varieties of cedar exist, and depending on where you live, cedar lumber
can sometimes be tricky to find. Western red cedar is the most widely
available type across the United States.

✔ **Spruce and fir.** Pine, spruce, and fir trees are all conifer trees. But when
it comes to nomenclature in the lumberyard, spruce and fir are typically
associated with *stud* lumber (versus *board* lumber). The plans in the book
make use of spruce or fir studs (either spruce or fir is fine, as they're interchangeable) that you use for making frames, top bars, and some other
applications.

Synthetic wood

Environmentally friendly, synthetic wood is made from a blend of recycled
plastics, sometimes combined with wood fibers. It's quite remarkable stuff.
It's completely weatherproof, will never rot, and is essentially maintenance-
free (just wash it clean). In recent years it has gained popularity as a very
durable decking material.

The jury's out when it comes to making beehives with this stuff. For one, it's
not available in all standard lumber sizes, so making a full-blown hive may be
a challenge. Also, it's astonishingly heavy and seriously expensive. Regarding
the designs in this book, I'd only consider synthetic wood for building hive
stands (see Chapter 13).

Shooting for the stars with exotic woods

If you want to knock someone's socks off with an over-the-top beehive or hive-top feeder, consider splurging and making one out of a high-grade or exotic wood. Cherry wood makes stunning kitchen cabinets — why not a cherry wood beehive? One year a friend and I made a couple of Langstroth hives using African mahogany. We sold them for $1,500 each! I also have a beautiful black walnut observation hive. It looks more like a fine piece of furniture than a beehive. The choice of wood is up to you. These fancy schmancy hives may not be any more functional, and perhaps even less practical, but if you're making hives to sell or just like having something unique, give it a go. Keep in mind, however, that your sweet bees don't know the difference between ponderosa pine and golden-grain Macassar ebony.

Woods to be wary of

Over the years I've heard other beekeepers talk about woods that may be toxic to bees and therefore shouldn't be used to make hives. One beekeeper told me that black walnut was one such wood. However, I have a black walnut observation hive and a colony that has thrived in it for years. As I researched this topic for the book, I found no hard evidence of a natural wood that has been proven toxic to honeybees. However, the sawdust created when working with some woods can be toxic or allergenic to the *woodworker* (examples are black walnut, mahogany, and cedar). But there's no evidence that these or other woods are problematic to the bees, so I wouldn't lose sleep over this. Chances are that any of the woods you can get your hands on are okay for making hives and equipment.

The one possible exception (and this is only my opinion) is pressure-treated wood. I just don't like exposing my bees to chemicals. Although since 2004, this category of wood product no longer uses toxic copper, chromium, and arsenic (CCA) to protect it from insects and mold. The new recipe is supposed to be safe. But I'd rather not take a chance with my girls. I'll stick to the all-natural, untreated wood, and I suggest you do so too.

Sizing up lumber

Most everyone has heard of a *two-by-four*. It's the lumber used to frame a house. And that's how you'd order it at the lumberyard: "I need a two-by-four." But did you know that it doesn't measure 2 inches by 4 inches? It actually measures 1½ inches by 3½ inches. However, when the board is first rough-sawn from the log, it *is* a true 2x4. By the time it's dried and planed, it becomes the finished 1½ inch by 3½ inch size.

Nominal dimensions refer to lumber *before* it's dried and planed to size at a lumber mill. You use nominal dimensions when you order wood. *Dimensional lumber* is the product that leaves the mill *after* drying and planing. These are the actual dimensions that you have to work with. I take all these measurements into consideration for the plans in this book. Throughout the building plan chapters, the materials lists refer to the nominal dimensions (how you order your materials), and the cut lists refer to the actual dimensional size (what you have to work with).

Table 3-1 lists the nominal and dimensional measurements for various sizes of lumber.

Table 3-1	Nominal Dimensions versus Actual Dimensions
Nominal Dimensions *(How You Order Lumber)*	**Dimensional Lumber Size** *(What You Have to Work with)*
⁵⁄₄ x 6 (decking)	1" x 5½"
1 x 3	¾" x 2½"
1 x 4	¾" x 3½"
1 x 5	¾" x 4½"
1 x 6	¾" x 5½"
1 x 8	¾" x 7½"
1 x 10	¾" x 9½"
1 x 12	¾" x 11½"
⁵⁄₄ x 3	1" x 2½"
2 x 3	1½" x 2½"
2 x 4	1½" x 3½"
2 x 6	1½" x 5½"
4 x 4	3½" x 3½"

Getting the scoop on plywood

There are many kinds of plywood out there. True plywood is made up of multiple thin layers of laminated wood veneers, each oriented at a 90-degree angle to the previous layer. This results in a product that resists warping and is incredibly strong. But note that any plywood meant for *interior* use won't last long in your bee yard. After a few rains or snowstorms, the layers that make up the plywood quickly peel apart like an onion. Here are the two types of plywood I use in this book:

✔ **Exterior plywood:** The plywood I specify for bottom boards and some other applications that are exposed to weather is either ¾ inch or ⅜ inch *exterior* grade plywood. Exterior plywood uses water-resistant glue to adhere its layers together. Don't use anything other than this for exterior applications.

✔ **Lauan plywood:** A number of the plans in the book specify using ¼ inch *lauan* plywood. The name *lauan* comes from trees found in the Philippines but has become a generic term in the United States for any imported tropical plywood. It's a lightweight product that's easy to work with and readily available. I've specified lauan plywood for those interior applications that are *not* exposed to the weather (such as inner covers and crown boards).

The Buzz on Other Building Materials

There's more to your beehives and beekeeping equipment than the lumber they're made from. You'll be dealing with all kinds of bits of hardware, plus various screws, nails, and staples as you assemble your hives and equipment. In this section I touch on each of these.

Fixating on fasteners

Fasteners are the screws, nails, and staples you use to assemble your hives and equipment. There are gazillions of different screws and nails in the marketplace. I've tried to make life as easy as possible by using only a few kinds of fasteners for the designs in this book. So if you plan to build a number of different things from these plans, you can stock up on fasteners and take advantage of some quantity discounts (the unit price of nails in a big box is far less expensive than the unit price of nails in a small convenience pack).

For fasteners that will be exposed to the exterior elements, I specify a *galvanized* product because these won't rust out or stain your woodenware. Epoxy-coated fasteners are also okay and seem to be gaining popularity. Alternatively, you can use stainless steel fasteners, but this option is way more expensive.

The types of screws used in this book include the following:

✔ #6 x ⅝ inch wood screws, #2 Phillips drive, flat-head (see Figure 3-12a)

✔ #8 x ½ inch lath screws, galvanized, #2 Phillips drive, flat-head with sharp point (see Figure 3-12b)

✔ #6 x 1⅝ inch deck screws, galvanized, #2 Phillips drive, flat-head with coarse thread and sharp point (see Figure 3-12c)

✔ #6 x 2½ inch deck screws, galvanized, #2 Phillips drive, flat-head with coarse thread and sharp point (see Figure 3-12d)

The types of nails used in this book include the following:

✔ 6d x 2 inch galvanized nails (see Figure 3-12e)

✔ ⁵⁄₃₂ inch x 1⅛ inch flat-head diamond-point wire nails (see Figure 3-12f)

✔ ⅝ inch finish brad nails (see Figure 3-12g)

✔ ⅜ inch staples for use in a heavy-duty staple gun (see Figure 3-12h)

The only other fastener I use is foundation pins for making frames (see Figure 3-12i), which are available from any of the major beekeeping supply houses.

Although drywall screws look really similar to deck screws and are much less expensive, don't be tempted to use them for your beehives. They are weaker and aren't treated to hold up to exposure to the elements. They'll rust and corrode in no time flat.

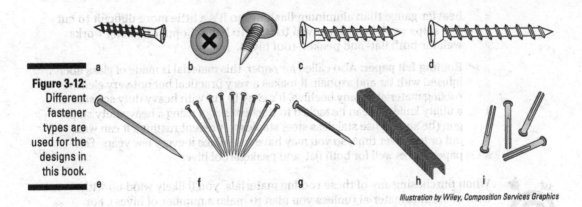

Figure 3-12:
Different
fastener
types are
used for the
designs in
this book.

a   b   c   d

e   f   g   h   i

Illustration by Wiley, Composition Services Graphics

Rifling through roofing materials

The roof of your hive takes the brunt of abuse from sun, rain, snow, and other climatic challenges. The roof (or outer cover) is a critical component for keeping your hive dry and your bees safe from the elements. You can optimize the roof's effectiveness and maximize its durability by adding some form of weatherproofing (just like the roof on your house). Most of the following options are interchangeable with one another. Pick the roofing material that best meets your needs and preferences.

✔ **Aluminum flashing:** The designs in this book specify aluminum flashing as a good roof material. This is the stuff that's used on the roofs of houses as a moisture barrier. It's easy to cut (with tin snips or even a utility knife) and a breeze to bend and shape. It's also readily available from any home improvement store. Aluminum flashing works well for both flat- and peaked-roof hives.

✔ **Asphalt roofing shingles:** This option is a little fussier because the application of roofing shingles takes a bit more work than some of these other options. But shingles are very practical and can give your hive a "little house" appearance. Roofing shingles come in a huge selection of colors. They work best on peaked-roof hives. Use a roofing nail that's short enough that it won't go all the way through both the shingles and the wooden hive top (¾ inch should be ample).

✔ **Cedar shake shingles:** Now we're talking! This option results in a very elegant cottage look. If you're selling your handiwork, this roof treatment would be very attractive to potential buyers. You won't find this kind of stylishness in your typical beekeeping supply catalogue. Roofing shingles work best on peaked-roof hives. Use a roofing nail that's short enough that it won't go all the way through both the shingles and the wooden hive top (¾ inch should be ample).

✔ **Copper flashing:** This is similar to aluminum flashing, but, as the name indicates, it's made of copper. It's pricey, but it can really dress up a hive if such aesthetics are important to you. Copper weathers to a soft green patina (like the Statue of Liberty). If you're building hives to sell, this extra bit of bling can be attractive to buyers and even command premium prices for your hives. Copper flashing tends to be available in a heavier gauge than aluminum flashing, so it's a little more difficult to cut (tin snips work best) and even trickier to bend. Copper flashing works well for both flat- and peaked-roof hives.

✔ **Roofing felt paper:** Also called *tar paper,* this material is made of glass fiber infused with tar and asphalt. It makes a very practical but not very elegant roofing material for any beehive. It's easy to cut with heavy duty scissors or a utility knife and can be tacked to the hive's roof using a heavy-duty staple gun (be sure to use stainless steel staples to prevent rusting). It can wear out or tear over time, so you may have to replace it every few years. Felt paper works well for both flat- and peaked-roof hives.

When purchasing any of these roofing materials, you'll likely wind up with a lot of surplus material (unless you plan to make a number of hives). For example, aluminum flashing is typically sold in a 10-foot roll, and you only need around 2 feet of flashing for the roof of one hive. So rather than purchase more than you need, visit a home construction site and see if the foreman is willing to give you some leftover material to cover the roof on your hive. You don't need that much for one hive. A jar of honey may just be the perfect barter for a couple of feet of flashing or a few dozen roof shingles!

Selecting screening materials

A few of the designs in the book call for screening material designed to keep the bees contained in a given area. But this isn't your ordinary window screening material (that stuff is too fine, and the bees would surely clog it up with propolis). Instead, the screening material of choice for beekeeping is *hardware cloth.* This consists of wire that's woven and welded into a grid. Specifically, you need hardware cloth with ⅛-inch square openings. It's known as #8 hardware cloth. It typically comes in 3-foot-by-10-foot rolls. If your local hardware store doesn't have #8 hardware cloth (they likely won't), you can easily find it online. Some beekeeping supply stores are now selling it by the foot (see www. bee-commerce.com or www.brushymountainbeefarm.com). You can cut hardware cloth to size using tin snips or even a pair of extra-heavy-duty scissors (see Chapter 4 for info on cutting and bending hardware cloth).

Protecting Your Beehives with Paint and More

After you build your masterpiece, you'll want to protect it from the elements. After all, most of this beekeeping stuff is intended to be kept outdoors year-round, where rain, sleet, and the gloom of night can take their toll. That's why I recommend that you apply some kind of protection to all exterior wooden surfaces. Protecting your woodenware with exterior grade paint, weatherproof polyurethane, or marine varnish not only looks nice but also adds many seasons of life to your wooden beehives and equipment.

Here are some options for protecting the exterior of hives:

✔ Most beekeepers paint their hives using a good quality exterior paint. Either latex or oil-based paints are fine. I find semi-gloss paints easier to keep clean than matte. As to color, it really doesn't matter — it's up to you. White is the most traditional color in the United States, but in other countries, the sky's the limit. Like your house, you'll need to repaint the hive every few years if you want it looking fresh.

Keep in mind that really dark colors heat up more than light colors. Dark hives may be fine in the cold winter but can quickly overheat the colony in the dog days of summer. So avoid black, navy blue, and aubergine!

✔ Another approach is to polyurethane your woodenware. You can even stain the wood first if you want to. This is an attractive, natural-looking alternative to paint. In fact, this approach not only protects the wood but also tends to blend in better with the environment (which can be helpful if you don't want to draw attention to your hives). Be sure to use exterior grade polyurethane; gloss, semigloss, or satin is fine. I prefer oil-based polyurethanes because they tend to stand up better to the elements. The resulting finish isn't very flexible and may crack over time and require touching up.

✔ Alternatively, you can apply several coats of marine-grade (spar) varnish to your woodenware. Again, you could stain the wood first if you want to. Spar varnish is slower drying but more flexible than polyurethane. Its flexibility means it tends not to crack. However, it must be reapplied periodically as it does tend to chip and flake off over time.

All the interior parts of the hive are best left *untreated* — no paint, no stain, no polyurethane, and no varnish. Keep the interior (where the bees live) au naturel. Doing so is healthier for the bees, and the bees prefer the natural living conditions because they're more like the inside of a tree, where bees live in nature.

Your hives and accessories undergo a good deal of stress over the seasons from weather conditions and from you frequently opening and closing the hives, moving the hives, swapping hive bodies, harvesting honey, and on and on. The goal is to build strong assemblies that stand the test of time. You can take a little extra step to ensure the wooden parts hold together for a long, long time. Throughout the book I suggest using a good-quality, weatherproof wood glue in addition to nail and screw fasteners. Though gluing isn't absolutely necessary, I can assure you your joinery will have a better chance of long-term survival if you take the time to apply a thin coat of weatherproof wood glue before screwing or nailing wooden parts together. Chapter 4 has more information on "going with glue."

Estimating the Amounts and Costs of Materials

Each building plan in this book includes a materials list that shows the exact amount of lumber, hardware, and fasteners you need for a given project. If you plan to build *two* of a particular hive design, you need to *double* what's in the materials list, and so on. If you plan to build several different projects in the book, make a shopping list by adding up the inventory from all the materials lists.

If you do plan on building more than one of the designs in this book or several copies of a particular design, I urge you to purchase some of your materials in bulk and thereby take advantage of quantity discounts. For example, many of the designs make use of the same size and type of deck screw. If you're building just one item, purchasing the exact quantity of deck screws you need in small-count blister packs makes sense. However, if you plan to churn out a bunch of stuff, save money by purchasing a larger quantity of deck screws. The same is true for the lumber and other materials specified in this book.

Each building plan also indicates what that particular design is likely to cost. These estimates are pretty rough because there are so many variables to consider. For example, where you live can have a big impact on the cost of goods. Lumber and hardware costs more in a large city than in a suburban or rural area. The big-box home improvement stores are likely to have better bargains than your neighborhood hardware store. Use my cost estimates as general comparative guidelines and worst-case scenarios. Beekeepers are a resourceful and clever tribe — we find all kinds of ways to minimize our out-of-pocket costs.

Using scrap inventory from previous projects is a terrific way to dramatically cut costs. Anything you don't have to buy is money in your pocket. As long as the materials are sound and free of harmful chemicals, you have no reason not to use them. Besides, beekeepers by their very nature are improvisers, and they're typically strong proponents of protecting the environment. So building beehives is a great opportunity to support your values. Do you have some scraps of lumber stacked in a corner of the basement? Maybe you have some leftover plywood from a previous project? Or how about that half-empty box of roofing shingles in the woodshed? If you're like me, you squirrel away these things, hoping that someday you may use them. Well, congratulations — that someday has arrived!

Chapter 4

Fine-Tuning Your Carpentry Skills

In This Chapter

- Making room for "bee space"
- Measuring, marking, and cutting lumber
- Understanding different joinery methods
- Handling flashing and wire
- Putting all the pieces together

Before you dive into building beehives and beekeeping equipment, draw a deep breath and take stock of your carpentry skills. Even if you have a workshop full of the best quality tools and finest lumber, that doesn't mean a hoot if your technique isn't up to snuff. In this chapter, I pass along some helpful tips for using your tools safely and provide information that will give you confidence as you embark on this hive-building adventure.

Note: I don't cover the particulars of building Langstroth-style frames in this chapter; they're complicated enough that most readers would rather buy them from a beekeeping supply vendor. That's fine! But if you're interested in trying to build them from scratch, flip to Chapter 17 for details on how to make and assemble your own frames.

Always Adhering to the "Bee Space"

You may wonder, why all the measuring fuss when it comes to beehives? Why can't you just put bees into any old box or container? In fact, you could put bees into almost anything, but that's assuming you don't plan on inspecting your colony or harvesting any honey.

Adhering to a measurement called the *bee space* lets you build hives that allow for the easy removal and inspection of combs and the simple separation and manipulation of the hive boxes.

The bee space is simply the crawl space that bees need to pass easily between two structures (⅜ inch is the ideal space). When the space between two surfaces in the hive is the right size (bee space), the bees will respect the space and leave it free as their passageway. But if the space between any two surfaces in a hive is much less than ⅜ inch (for example, less than ¼ inch), the bees will quickly seal the gap with sticky *propolis* (a resin-like substance the bees manufacture to seal cracks and gaps in the hive). And if the space is much larger than ⅜ inch (for example, more than ½ inch), the bees will fill the space with extra wax comb. Either way, violating the bee space results in a colony that you can't easily inspect and manipulate because the bees have effectively glued everything together!

You want to adhere to this bee-space concept as you build your hives (in other words, the gaps between surfaces are never more or less than ⅜ inch). The good news is that I've done all the calculations for you! Every plan in this book allows for the correct bee space. Just adhere to the measurements in the cut lists and all will be well.

Two types of bee space exist. The ⅜ inch bee space your colony uses to move between hive bodies can be designated either at the top or bottom of each hive body.

- Top bee space means the passageways between hive bodies have been designated *above* the frames.
- Bottom bee space means the passageways between hive bodies have been designated *below* the frames.

Although beekeepers endlessly debate whether top or bottom bee space is better, you can't mix top space and bottom space equipment in the same hive. Top bar configuration is standard in the United States and many other countries around the world. It's even gaining favor in the UK (where bottom bee space has been traditional). So the plans for *all* the hives in this book use *top* bee space (even the decidedly British hive that's called, naturally, the British National hive).

If you decide to purchase equipment from a beekeeping supply store to use with the hives you build (such as frames, hive-top feeders, and so on), be sure to confirm with the vendor that the products you're buying are compatible with *top* bee space.

Measuring and Marking Lumber

Figuring out how to read a tape measure is one of the most vital skills for building a beehive. Take a moment before you start cutting anything to become familiar with your tape measure and how to use it. How well can you decipher those little marks on your tape measure (see Figure 4-1)? Do you know which of those are eighths versus sixteenths? Some tape measures even show thirty-seconds of an inch (time to get out the reading glasses!).

Note: In an effort to keep things as simple as possible, very few of the measurements in this book are less than ⅛ inch increments.

After you measure lumber carefully, place your pencil exactly on the spot to be cut. Then bring your carpenter's square into contact with the pencil tip. Now scribe a long, easy-to-see line across the entire width of the board. This little trick of bringing the square in contact with the pencil ensures that the line remains at the measured distance versus being off by the thickness of the pencil.

Figure 4-1:
This particular tape measure design is perfect for the plans in this book because the markings include helpful identification of key fractions of an inch.

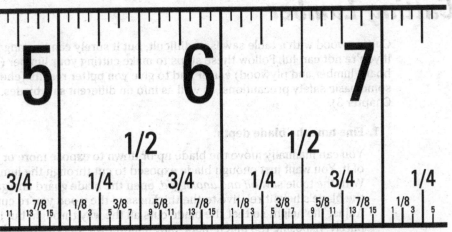

Illustration by Wiley, Composition Services Graphics

"Measure twice, cut once" is an old saying everyone has heard many times — and for a good reason. It's terrific advice. After your lumber is cut, it's irreversible. Take the time to double-check every measurement for accuracy before you rev up your saw.

Now, before you make the cut, scribble an X onto the side of the board you won't be using (see Figure 4-2). That's the scrap piece. Why mark the scrap, you ask? Well, say you have a 48-inch board and you need a piece that's exactly 24¼ inches long. You measure it and make your cut. As you finish the cut, both pieces of lumber fall to the ground. Drat! Both pieces look to be the same length, but they're not. Which piece is the piece you need and which is scrap? Putting an X on the piece you *don't* need prevents this dilemma. Neat trick, huh?

Figure 4-2:
Marking where the cut should be made and which is the scrap piece assures accuracy and avoids mix-ups.

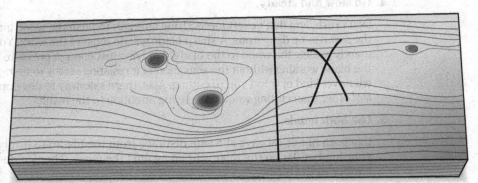

Illustration by Wiley, Composition Services Graphics

Cutting Lumber

Cutting wood with a table saw isn't difficult, but it surely can be dangerous if you're not careful. Follow these steps to make cutting your lumber (both board lumber and plywood) easier and to give you better results (check out some basic safety precautions, as well as info on different saw blades, in Chapter 3).

1. **Fine-tune the blade depth.**

   You can manually move the blade up or down to expose more or less of it. You want just enough blade exposed to cut through the lumber. With the table saw *off and unplugged,* open the blade guard and gauge the blade's height relative to the thickness of the wood you're cutting. Adjust the height so that the blade cuts all the way through, but just barely. Exposing too much blade may splinter the wood rather than make a crisp, clean cut.

2. **Start the blade early.**

   Place the lumber you intend to cut on the table saw. The blade shouldn't be touching the wood when you fire up the table saw. Allow the saw to run a few seconds to get to full speed before beginning your cut. Doing so is easier on your equipment and also prevents the wood from splintering.

3. **Keep the kerf in mind.**

   The *kerf* is the miniscule amount of wood that's removed by the width of the blade itself (usually around ⅛ inch). You want to take the kerf into account as you line up the blade with the measurement mark you so carefully penciled on the board. And do you remember that X you put on the scrap side of the lumber (see the preceding section)? Position the blade on the same side of the measurement line as the X. If you "erase" that line with the cut, your piece may be a fraction of an inch less than you intended.

4. **Go slow and steady.**

   Don't rush, and don't exert force. Using strength to push the wood into the saw can be dangerous. A sharp blade will do most of the work. You're merely *guiding* the wood to the blade. Applying just enough pressure to feel the blade working without the sensation of it resisting seems to be right for whatever kind of wood you're cutting. Alas, there's no way to develop that *feel* other than cutting wood and paying attention to the results.

5. **Use both hands.**

   Use both hands (always at a safe distance from the blade) to firmly hold your lumber flat on the table and up against the fence. Then guide the lumber to the blade.

A push stick is a *must* when working with a table saw. It keeps your hands away from the spinning blade as you feed the wood closer and closer to the saw. Although you can purchase a commercially made push stick with extra bells and whistles, practical ones are easy to make. Use a piece of ¾-inch-thick scrap wood cut to fit comfortably in your hand. A notch at the business end of the stick holds the wood as you push it toward the saw (see Figure 4-3). A simple hand-held jigsaw comes in handy for making something like this.

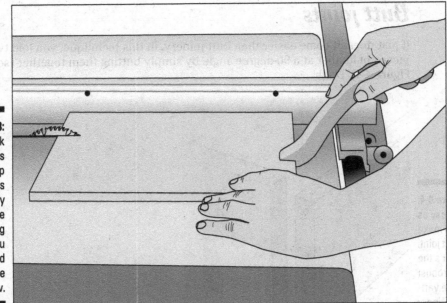

Figure 4-3:
A push stick jig allows you to keep your fingers safely away from the spinning blade as you feed wood through the table saw.

Illustration by Wiley, Composition Services Graphics

A *saw kerf cut* is the incision made by a saw in a piece of wood. You can use a kerf cut to create a groove or slot in a piece of wood. The width of the resulting cut is equal to the thickness of the saw blade itself (typically ⅛ inch). A number of the plans in the book specify making a kerf cut. In these cases, you don't cut all the way through the wood. You simply cut a groove/slot that's ⅛ inch wide by the depth specified in the plans.

Plywood tends to splinter easily, even when you use a saw blade intended for plywood. Here's a way to minimize splintering: After you mark your cut line with a pencil, use a utility knife to score along the mark. Use a steel straightedge and make a few passes with the utility knife. This breaks the fibers in the topmost layer of plywood and reduces splintering when you make the cut with the table saw.

What's Up with This Joint?

You can join wood together using all kinds of techniques. Collectively, these techniques are referred to as *joinery*. The list of different joinery methods used in woodworking is long and diverse. Some techniques are very complex and require special skills and equipment. The good news is that, in this book, I keep it simple. You use only four different joinery techniques for the builds, as I explain in the following sections.

Butt joints

It just doesn't come easier than butt joinery. In this technique, you join two pieces of lumber at a 90-degree angle by simply butting them together (see Figure 4-4). Done!

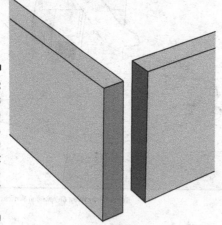

Figure 4-4: It's easy as pie to make a butt joint, but it's the least robust of the various joinery techniques.

Illustration by Wiley, Composition Services Graphics

WARNING! Although the butt joint is the simplest to make, it's also the weakest. And over time, the rain, humidity, heat, and cold cause the joinery to split open. Though these openings may provide some nice ventilation and an extra entrance for the bees, you'll find them quite impractical in every other way. Make this simple but weak joint as strong as possible by cutting the wood true and clean and by using a weatherproof wood glue and screws (not nails).

Rabbet cuts and dado joints

It took me a while to sort out the difference between rabbets and dadoes. Here's a definition of each of these similar joinery techniques:

- **Rabbet:** I'm not talking bunnies here. Those are *rabbits*. A *rabbet* (called a *rebate* in the UK) is a recess cut into the *edge* of a piece of wood. When viewed in cross-section, a rabbet is two-sided and open along the edge of the wood. Think of a rabbet as the letter *L* (see Figure 4-5). Rabbet cuts are sometimes used to join two pieces of wood together, but they're also used to create shelves or ledges (such as the shelf upon which the frames rest).

- **Dado:** I'm not talking defunct birds here. Those were *dodos*. A *dado* (called a *housing* or *trench* joint in Europe) is a grooved slot cut into the surface of a piece of wood. When viewed in cross-section, a dado has three sides. Think of a dado as the letter *U* (see Figure 4-6).

The measurement of the dado cut is identical to the thickness of the piece of wood that fits into the dado. For example, you use this form of joinery when building a bottom board. A ¾-inch dado cut into the side rails accommodates the ¾-inch-thick plywood floor. Dado joinery results in a solid, strong connection between two pieces of wood.

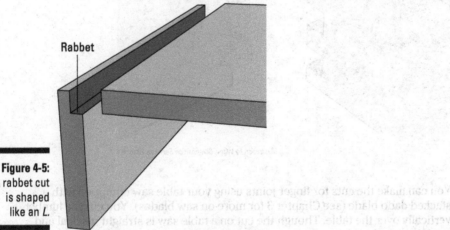

Rabbet

Figure 4-5:
A rabbet cut
is shaped
like an *L*.

Illustration by Wiley, Composition Services Graphics

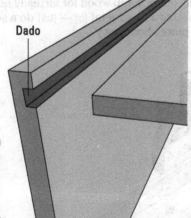

Dado

Figure 4-6:
A dado
is shaped
like a *U*.

Illustration by Wiley, Composition Services Graphics

Finger joints

Finger joinery (also known as *box joinery* or *comb joinery*) involves square interlocking fingers that join two pieces of lumber at a right angle (see Figure 4-7). Although a little tricky to make, this is, hands down, about the strongest method of joining two pieces of wood together.

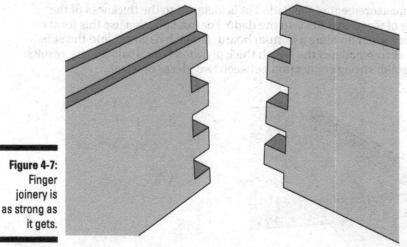

Figure 4-7:
Finger
joinery is
as strong as
it gets.

You can make the cuts for finger joints using your table saw equipped with a stacked dado blade (see Chapter 3 for more on saw blades). You cut the lumber vertically over the table. Though the cut on a table saw is straight vertical and very precise, the tricky part is to get the spacing right. Here's where box joint jigs or templates come into play. A jig saves time and guarantees the precision of your joints. You can purchase a commercially made one for around $60 (see Figure 4-8). Or you can make your own jig out of scrap wood for virtually nothing. The Internet has tons of plans for making a finger joint jig — just do a search on the web. (See the nearby sidebar for more about jigs.)

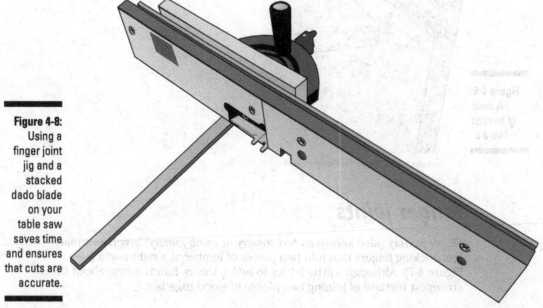

Figure 4-8:
Using a
finger joint
jig and a
stacked
dado blade
on your
table saw
saves time
and ensures
that cuts are
accurate.

Doing a jig

Woodworking jigs are like having an extra set of hands. Jigs allow you to do things with a tool (a table saw, for example) that make the task safer or more accurate (or both). When it comes to building beehives, you can use a jig to control the placement and/or motion of another tool. An everyday example is when a key is duplicated; the original key is used as a jig. When cutting the new key, the blade simply follows the same profile as the old key. When used to control another tool, a jig's purpose is to provide repeatability and accuracy when cutting your woodenware.

You can also use a jig as an aid when making multiple repetitions of the same work; this kind of jig is sometimes referred to as a *fixture*. It holds your work piece so that both of your hands are free to work on the piece.

All in all, jigs make life easier and safer for you when you're cutting and assembling your woodenware. Here are a couple of different jigs to consider for the projects in this book:

✔ **Jig for making finger joints:** The hive bodies and honey super for the Langstroth hive (see Chapter 10) use a finger joint. Though you don't have to use this form of joinery (you can use a simpler butt joint or rabbet joint), finger joints are traditional for this hive, and they're by far the strongest joinery technique for building beehives. You can purchase a finger joint jig online or at most woodworking supply stores, or you can make your own from scrap wood.

✔ **Jig for assembling frames:** Whether you make your own frames from scratch (see Chapter 17) or purchase frames from a bee-keeping supplier, you still have to assemble, glue, and nail the parts together. At best it's a monotonous chore. But you can make the task much easier (and certainly faster) using a frame jig (or frame fixture). Not all beekeeping supply stores carry these, so now you can make your very own (see Chapter 11). The jig essentially holds all the parts to assemble and nail ten frames at a time. It's so much easier when you don't have to juggle parts and nail frames one frame at a time.

Working with Flashing and Wire

Wood isn't the only material you work with as you build beehives. A number of the designs in this book make use of metal — specifically, aluminum flashing and wire hardware cloth. The following sections tell you what you need to know.

Cutting and bending metal flashing

I specify aluminum flashing as a roof material for some of the hives in this book. It's very thin and easy to cut and manipulate. This is the same material that roofers use to waterproof critical seams. Any big-box home improvement store carries aluminum flashing. You need 20-inch-width material, which likely comes in a 10-foot roll. You can use the extra material when building four additional hives.

The edges of metal flashing are very sharp. Use caution when handling flashing to avoid cutting yourself, and consider using work gloves.

Cutting flashing is easy. You have a couple of options:

✔ You can breeze through it using a pair of tin snips (see Chapter 3); be sure to measure and mark it carefully before you make a cut. To make your mark, you can scratch the metal with a nail or use a felt-tip marker. Note that the snips do tend to curl the cut edges slightly, but that's not a big deal.

✔ Alternatively, you can use a sharp utility knife and a straightedge to cut the flashing. With your flashing lying flat on your worktable, measure and mark where you plan to make your cut. Lay down your straightedge on the mark and make several authoritative passes with the utility knife. Done!

Aluminum flashing is so thin, it's easy to bend. You don't need special equipment. But for crisper bends and folds, bend the flashing over the 90-degree edge of a table or the edge of a sheet of lumber.

If you trade up and decide to use 40-ounce copper flashing on your beehives, this bending is best done using a special machine called a *sheet metal brake*. The metal brake allows you to make perfect folds when using heavier material (such as copper flashing). You can find these online, at big-box home improvement stores, and at roofing supply vendors.

Cutting and shaping wire hardware cloth

Hardware cloth consists of wire that's woven and welded into a grid. You use it to keep the bees from traveling from one part of the hive to another. Here's the thing: Hardware cloth is very easy to find with either ¼-inch or ½-inch square openings in the mesh, but those holes are too large, and the bees will breeze right through. What you need is hardware cloth with ⅛-inch square openings. It's known as #8 hardware cloth (see Figure 4-9). It typically comes in 3-foot-by-10-foot rolls. If your local hardware store doesn't have #8 hardware cloth, you can easily find it online. Some beekeeping supply stores sell it by the foot (see www.bee-commerce.com or www.brushymountainbeefarm.com).

Figure 4-9: The hardware cloth you purchase must have ⅛-inch openings in the grid. That ensures it's "bee-tight."

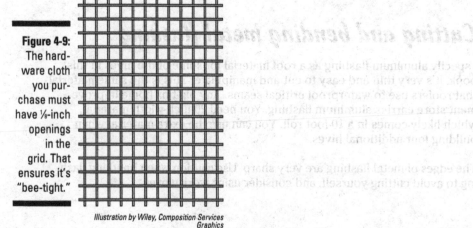

Illustration by Wiley, Composition Services Graphics

Cutting hardware cloth is easy using tin snips. Or you can use heavy-duty scissors. Make the cut as close to one of the vertical wires as possible to avoid having ragged horizontal wires sticking out of the side. Ouch! Use a felt-tip marker to measure and mark the size you need to cut.

The hive designs that call for hardware cloth don't call for much bending. But the hive-top feeder in Chapter 15 requires you to fold a couple pieces of hardware cloth. For this simple bend, place a wooden board at the location where the bend should be and pull the hardware cloth upward against the board to the desired angle.

Assembling the Parts of Your Hive

After you measure, mark, cut, and join all your materials, it's time to assemble everything into an honest-to-goodness beehive. The following sections provide a few simple hints that makes building your hives and equipment easier.

Going with glue

Throughout this book I suggest the use of glue in addition to nails and screws. And though doing so isn't mandatory, supplementing fasteners with weatherproof yellow carpenter's glue surely makes your hives and equipment as strong and long-lasting as possible.

For best results, before using your fasteners, wipe or brush a thin, even coat on both surfaces of the joint. Don't goop it on so much that it oozes out of the joint when the pieces come together. Work fairly quickly, as the glue will start to set up in a few minutes. Carpenter's glue cleans up nicely while it's still wet, so use a damp rag to wipe down the outside of the joints for a tidier look. After the glue dries it's the devil to remove, and if you plan to stain your woodenware, the stain won't take to those areas covered with dry glue.

Being square

When it comes to building beehives and beekeeping equipment, you're generally making boxes and frames, so it's really hip to be square. When two pieces of wood come together at right angles, you can't simply assume that they'll meet at a perfect 90-degree angle. If that angle is just a wee bit off and you keep nailing things together, your project will get more and more out of whack the further you go.

Here's where your carpenter's square comes into use (see Chapter 3). Use a square to check how your pieces fit together before you use any nails and screws, and then periodically check your assembly with your square to visually confirm that your hive bodies, supers, frames, and other parts are joined at the perfect 90-degree angle.

Nailing and screwing everything together

I'll bet a hammer was one of the first tools you ever used as a kid. But surprisingly, many weekend warriors haven't mastered a good technique for using a hammer. Here's a tip: Don't be a wimp. Get a good grip at the end of the handle (see Figure 4-10) — not the middle and not up by the head. Use an authoritative swinging motion from the elbow and hit the nail with the center of the hammer's face. Tapping lightly is okay at first to get the nail set, but after that you should be able to drive the nail home with just a few committed swings.

Figure 4-10:
Grip a hammer at the end of the handle and use your whole forearm to swing.

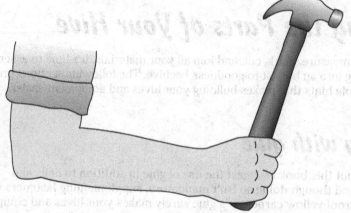

Illustration by Wiley, Composition Services Graphics

Eliminate bashed thumbs while setting nails — use a cheap plastic comb to hold a nail in position while you tap away.

For the screened bottom board in Chapter 14, I specify *toe-nailing* two pieces of wood together. Toe nailing has nothing to do with the digits on your feet. It's a technique of joining two pieces of wood by driving a nail at an angle. It's easiest done when the two parts to assemble are braced against some kind of stop on your worktable. A piece of lumber secured to the work surface makes a good stop. Place the nail at a 45-degree angle and tap it into place.

When it comes to using screw fasteners, I've made life easy for you. All the screw fasteners used in this book are specified with #2 Phillips heads. That means the only screw bit you need to use is a #2 Phillips. Rather use a hex or square-head type of screw? No problem. It's up to you.

As I assemble various hive parts, I like to first test-fit the wooden pieces (making sure everything is okay), and then I apply a single fastener (screw or nail) into each side. I drive the nail or screw only partway in, just enough to hold the pieces together. This allows a little wiggle room to make certain everything is square. After I confirm that all is well, I drive the fasteners in all the way and add the remaining ones.

Part II

The World's Most Popular Beehive Designs

The 5th Wave   By Rich Tennant

"I don't know why you build those things when your brother-in-law just smashes them open as soon as they're finished and steals the honey."

Part II
The World's Most
Popular Beehive
Designs

In this part . . .

Now the fun really begins. Over the centuries, beekeepers have used myriad hive designs. In this part, you find plans for building six of the world's most popular beehive designs, from simple to complex. You can't go wrong with any of them! Each chapter includes a detailed list of the materials you need (your shopping list), an illustrated cut list that shows you how to trim your lumber and other goods, and detailed assembly instructions to help you put it all together.

Chapter 5

The Kenya Top Bar Hive

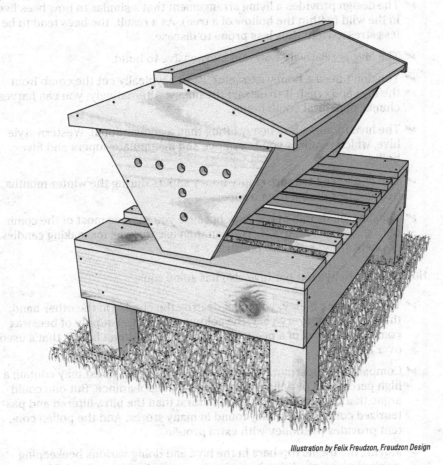

Illustration by Felix Freudzon, Freudzon Design

The top bar hive is likely one of the world's oldest beehive designs, going back centuries. In recent years, this simple, practical design has seen a resurgence among backyard beekeepers, particularly because more and more backyard beekeepers are seeking more natural beekeeping options. Top bar hives are used extensively throughout Africa and in other developing countries where building materials are scarce. The design in this chapter is called a *Kenya top bar hive;* its sloped sides tend to result in stronger combs and to discourage the bees from attaching the comb to the bottom of the hive.

The Kenya top bar hive has many design variations, but they all consist of a long, horizontal hive body with sloped sides, top bars (instead of frames), and a roof. The design in this chapter was developed by friend and fellow beekeeper Mike Paoletto. It's a very practical and functional hive, but like all top bar hives, it has some pros and cons. First the pros:

- The top bars (which, unlike frames, have no bottom or side rails and use no full sheets of foundation) allow the bees to build their comb freestyle and without restrictions on cell size.

- The design provides a living arrangement that's similar to how bees live in the wild (within the hollow of a tree). As a result, the bees tend to be less stressed and thus less prone to disease.

- The hive is relatively easy and inexpensive to build.

- You don't need a honey extractor, as you typically cut the comb from the hive and crush it to extract the honey. Alternatively, you can harvest chunks of natural comb honey.

- The hive requires less heavy lifting than a conventional, Western-style hive, which requires you to remove and manipulate supers and hive bodies.

- You don't need to store empty honey supers during the winter months because this design has no supers.

- The hive yields lots of beeswax because you remove most of the comb during the honey harvest. You can then use the wax for making candles, cosmetics, and furniture polish.

However, the Kenya top bar hive also has some cons:

- To harvest the honey, you must destroy the comb. On the other hand, that means that every year, the colony uses a fresh supply of beeswax comb, so chances of a pesticide buildup are less than in wax that's used over and over.

- Compared to supermarket honey, the honey you harvest may contain a high percentage of pollen and be cloudy in appearance. But one could argue that this honey is far more natural than the ultra-filtered and pasteurized commercial honey found in many stores. And the pollen content provides the honey with extra protein.

- Moving the fragile top bars in the hive and doing various beekeeping manipulations are difficult or even impossible compared to what's possible with a hive using Langstroth-style frames (which support the comb on all four sides). But the whole idea of this design is to provide as natural an environment as possible for your bees with fewer interruptions from the beekeeper. Those manipulations are contrary to the all-natural objective.

- Administering medications to your bees can be difficult (the hive doesn't have a feeder, for example). But hey, a top bar hive is all about a sustainable, more natural approach to beekeeping. So medication and chemicals need not apply. Bees tend to be healthier when under less stress and in a more natural environment, and this simple design provides just that.

✔ This type of hive has the potential of lower honey production because it has a fixed number of top bars versus more modern designs, where multiple honey supers and frames can be stacked endlessly, one on top of the other.

✔ Combs are much more fragile than those built in a Western-style frame (such as a Langstroth hive; see Chapter 10). You need to be very careful when inspecting combs of a top bar hive.

✔ Managing this hive requires a degree of special knowledge that's not typically found in conventional beekeeping books. However, the Internet has a growing number of useful articles and videos that are helpful to beekeepers interested in trying this time-tested design. Just do a search for "managing top bar hives."

✔ I've seen some chatter on the Internet stating that horizontal top bar hives don't winter well. Not true. I know a number of beekeepers who are successfully using this design here in New England (where we have plenty of cold, cold winter).

Vital Stats

✔ **Size:** 38 inches x 24 inches x 31¼ inches.

✔ **Capacity:** You can't add supers or additional hive bodies. The space is fixed and thus limited.

✔ **Type of frame:** This design uses a top bar, not a frame. It doesn't have side bars, a bottom bar, or full sheets of beeswax foundation to deal with (nor the costs associated with these elements). The bees build their comb naturally and without restriction onto each of the top bars placed in the hive (28 in all).

✔ **Universality:** All the gadgets and add-ons that you might use with a conventional Langstroth hive (feeders, queen excluders, frames, foundation, and honey-extracting equipment) are irrelevant to the Kenya top bar hive. Its design is virtually all-inclusive and requires no extras. And with no standardization of design, commercial replacement parts for top bar hives aren't available.

✔ **Degree of difficulty:** This is likely one of the easiest hives to build. It has no complex joinery and requires no complicated frame building or foundation insertion. That's why this design has remained so popular in Africa and other developing countries.

✔ **Cost:** Using scrap wood (if you can find some) would keep material costs of this design very minimal, but even if you purchase the recommended knotty pine, hardware, and fasteners, you can likely build this hive (including the top bars and the stand) for around $120.

Materials List

The following table lists what you'll use to build your Kenya hive and the top bars used with it. In most cases, you can make substitutions as needed or desired. (See Chapter 3 for help choosing alternate materials.)

Lumber	Hardware	Fasteners
2 8' lengths of 1" x 6" clear pine lumber	Weatherproof wood glue	20 #6 x 2½" deck screws, galvanized, #2 Phillips drive, flat-head with coarse thread and sharp point
1 8' length of 1" x 12" knotty pine lumber	⅛" hardware cloth (you'll need a piece that measures 36½" x 7¾"). #8 hardware cloth typically comes in a 3' x 10' roll, but some commercial beekeeping supply vendors sell it by the foot.	80 #6 x 1¾" galvanized deck screws with coarse thread and sharp point
1 8' length of 4" x 4" cedar posts	½ pound beeswax (to melt and brush on the frames' starter strips)	25 ⅜" staples for use in a heavy-duty staple gun
2 12" x 36" x ³⁄₃₂" balsa wood sheets	Optional: 2 quarts latex or oil exterior paint (white or any light color), exterior polyurethane, or marine varnish	

Here are a few notes about the materials for your Kenya top bar hive:

- Because of its simple butt-joinery, knotty pine is perfectly okay for this hive, and it's about the least expensive board lumber out there. But pine isn't the most durable wood; it benefits from a protective coating of paint, varnish, or polyurethane. Alternatively, consider keeping your hive as all-natural as possible by using cedar or cypress. Although these woods are more expensive than knotty pine, they stand up well to weather without paint or protective chemicals.

- You can purchase ³⁄₃₂ inch thick sheets of balsa wood online or from a hobby store. Use a utility knife to cut the balsa wood sheets into the 13⅝ inch x ¾ inch starter strips.

- I've tossed in few more fasteners than you'll use because, if you're like me, you'll lose or bend a few along the way. It's better to have a few extras on hand and save yourself another trip to the hardware store.

Cut List

The following sections break down the Kenya top bar hive into its individual components and provide instructions on how to cut and build those components.

Lumber in a store is identified by its *nominal* size, which is its rough dimension before it's trimmed and sanded to its finished size at the lumber mill. The actual finished dimensions are always slightly different from the nominal dimensions. For example, what a lumberyard calls *1 inch x 6 inch lumber* is in fact ¾ inch x 5½ inch, and 4 inch x 4 inch posts are actually 3½ inch x 3½ inch. In the following tables, each Material column lists nominal dimensions, and each Dimensions column lists the actual, final measurements.

You make a number of special cuts in this chapter, including rabbet cuts and saw kerfs. See Chapter 4 for detailed instructions on making these cuts.

You can adjust the height of the elevated hive stand to suit your needs by adjusting the length of the 4 inch x 4 inch cedar posts. Longer legs result in less bending over during inspections. I find the 13 inch height of the stand just right for me.

Elevated hive stand

Quantity	Material	Dimensions	Notes
4	4" x 4" cedar posts	12½" x 3½" x 3½"	These are the leg posts of the stand. Rabbet 5½" x ¾" deep along one end of each post (this rabbet accommodates the long sides of the stand).
4	1" x 6" clear pine	24" x 5½" x ¾"	Two of these are the short sides of the stand and the other two are the wide struts for the top.
2	1" x 6" clear pine	36½" x 5½" x ¾"	These are the long sides of the stand.
5	1" x 6" clear pine	24" x 2¼" x ¾"	These are the narrow struts for the top.

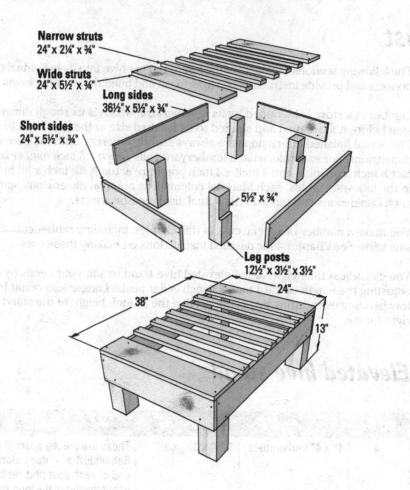

Narrow struts
24" x 2¼" x ¾"

Wide struts
24" x 5½" x ¾"

Long sides
36½" x 5½" x ¾"

Short sides
24" x 5½" x ¾"

5½" x ¾"

Leg posts
12½" x 3½" x 3½"

38"   24"   13"

Illustration by Felix Freudzon, Freudzon Design

Hive body

Quantity	Material	Dimensions	Notes
2	1" x 12" knotty pine	34½" x 10⅜" x ¾"	These are the long sides. Bevel the top long edge so that the outside height measurement of the board is 9" while the inside height measurement remains at 10⅜". This provides the necessary slope to accommodate the inclined roof boards.
2	1" x 12" knotty pine	18" x 11¼" x ¾"	These are the V-shaped end panels. The top edge is 18" and the bottom edge is 5⅛". These dimensions are centered, and they're what determine the V shape of this piece. Drill seven ¾" ventilation/access holes into *one* of the end panels (see the following figures for the approximate placement of the holes).
1	#8 hardware cloth	36½" x 7¾"	This is the screened bottom of the hive; staple it in place as indicated in the drawing.

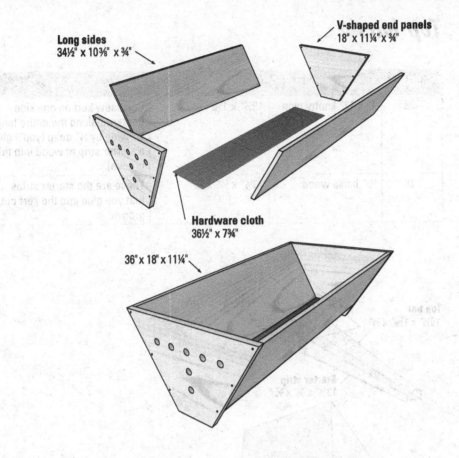

Long sides
34½" x 10⅜" x ¾"

V-shaped end panels
18" x 11¼" x ¾"

Hardware cloth
36½" x 7¾"

36" x 18" x 11¼"

Illustration by Felix Freudzon, Freudzon Design

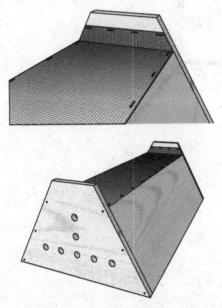

Illustration by Felix Freudzon, Freudzon Design

Top bars

Quantity	Material	Dimensions	Notes
28	1" x 12" knotty pine	19⅛" x 1⁵⁄₁₆" x ¾"	Cut a saw kerf on one side centered along the entire length, ⅛" wide by ¼" deep (you'll glue a starter strip of wood into this groove).
28	³⁄₃₂" balsa wood	13⅝" x ¾" x ³⁄₃₂"	These are the starter strips that you glue into the kerf cut groove.

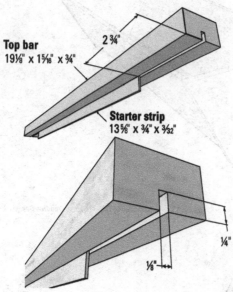

Top bar
19⅛" x 1⁵⁄₁₆" x ¾"

2 ¾"

Starter strip
13⅝" x ¾" x ³⁄₃₂"

¼"

⅛"

Illustration by Felix Freudzon, Freudzon Design

Ventilated roof

Quantity	Material	Dimensions	Notes
2	1" x 12" knotty pine	36½" x 10⅛" x ¾"	These are the inclined roof boards.
2	1" x 12" knotty pine	22" x 4¾" x ¾"	These are the roof gables. Cut into a V shape by leaving a 1" rise at the bottom and a 1⅝" flat cap at the top (see the following figures for details).
1	1" x 12" knotty pine	40" x 1¾" x ¾"	This is the roof ridge.
1	1" x 12" knotty pine	6¼" x 2¾" x ¾"	This is the roof support wedge. Cut into a V shape by leaving a 1⅝" rise at the bottom (see the figures for details).

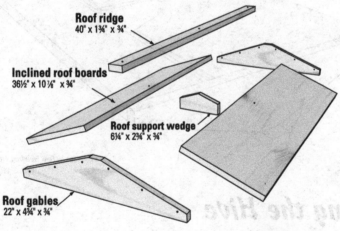

Roof ridge
40" x 1¾" x ¾"

Inclined roof boards
36½" x 10⅛" x ¾"

Roof support wedge
6¼" x 2¾" x ¾"

Roof gables
22" x 4¾" x ¾"

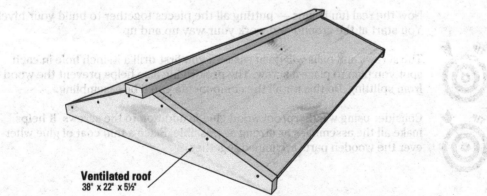

Ventilated roof
38" x 22" x 5½"

Illustration by Felix Freudzon, Freudzon Design

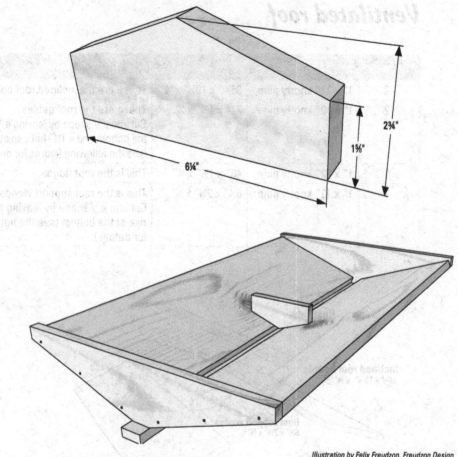

6¼"

2¾"

1⅝"

Illustration by Felix Freudzon, Freudzon Design

Assembling the Hive

Now the real fun begins — putting all the pieces together to build your hive!
You start at the ground and work your way up and up.

The screws and nails will go in easier if you first drill a ⁷⁄₆₄-inch hole in each
spot you plan to place a screw. The pre-drilling also helps prevent the wood
from splitting. Do this for all the components you'll be assembling.

Consider using weatherproof wood glue in addition to the screws. It helps
make all the assemblies as strong as possible. Place a thin coat of glue wherever the wooden parts are joined together.

1. **Assemble the hive stand.**

   Fasten the long sides into the rabbet cut of the leg posts and secure to each post using two 2½ inch deck screws per post. The edge of the side rail should be flush with the post. Stagger the placement of the screws (as shown in the earlier figure) to prevent splitting the wood.

   Using deck screws, attach the two short sides to the leg posts and the long sides. One screw goes into the leg post, and another goes into the edge of the long side rail. Use two screws for each corner.

   Attach the two wide struts to the top of the stand. Position each flush with the front and rear ends of the stand. Secure each end of each strut using two deck screws, as shown in the earlier figure.

   Attach the five narrow struts to the top of the stand. By eye, evenly space them between the two wide struts and secure them to the top edge of the long sides (use one deck screw at the end of each strut).

   Optional: Paint, varnish, or polyurethane the entire hive stand to protect it from the elements. Use two or three coats, letting each coat dry completely before adding the next coat. If you elect to paint the stand, any color will do — it's up to you. (See Chapter 3 for information on protecting your woodenware.)

2. **Assemble the hive body.**

   Affix the two V-shaped end panels to the long sides using 1⅜ inch deck screws (as shown in the earlier figure). Use a total of six screws on each end panel. Note that the top edges of the end panels and side boards should be flush with each other.

   Now turn the hive body upside down and staple the hardware cloth to cover the opening that runs along the entire length of the hive body. Place one ⅜ inch staple every couple of inches. At each end of the screening, bend the screening at a 90-degree angle and staple to the end panels. Trim excess screening as necessary using tin snips.

   Optional: Paint the *exposed* exterior wood of the hive body with a good quality outdoor paint (latex or oil). This greatly extends the life of your woodenware. You can use any color you want, but a light pastel or white is best. With dark colors, the hive builds up lots of heat in the summer, and your bees spend a lot of energy cooling the hive — energy they could be spending collecting nectar. Alternatively, you can use a few protective coats of polyurethane or marine varnish on the exterior wood.

3. **Assemble the top bars.**

   You'll assemble a total of 28 top bars. Assembly simply consists of gluing a thin strip of wood into the kerf cut groove. This is the starter strip that gives the bees a starting point to build honey comb on the top bar. For each top bar, center the starter strip into the kerf cut and glue in place using weatherproof wood glue. Let the glue dry before proceeding.

Melt ½ pound of beeswax over low electric heat or in a double boiler. Use a disposable brush to coat the starter strips with a thin coat of beeswax. This further encourages the bees to get started making comb.

Never melt beeswax using an open flame! Beeswax is highly flammable.

Now place the bars into the hive body. The bars rest on the top edge of the hive body and are butted side by side, like the wooden bars of a marimba (but with no gaps between each bar).

Never paint your top bars because that could be toxic to your bees. Leave all interior parts of any hive unpainted, unvarnished, and all-natural.

4. Assemble the ventilated roof.

Affix the two inclined roof boards to the two peaked gables using 1⅝ inch deck screws. The screws go through the peaked gables and into the edges of the roof boards. Use six evenly spaced screws per gable (see the figure for approximate placement; precise spacing isn't critical). Note that there's a 1⅝-inch-wide ventilation opening at the peak of the roof.

Now take the ridge rail and screw it in place using one 1⅝ inch screw at each end. Screws go through the ridge rail and into the flat top of the gable pieces.

Center the roof support wedge in the middle of the ventilation opening (underside of the roof) and secure in place with a 1⅝ inch screw driven through the ridge rail and into the support wedge.

Optional: Paint the *exposed* exterior wood of the roof assembly with a good quality outdoor paint (latex or oil). You can use any color you want, but a light pastel or white is best. Alternatively, you can use a few protective coats of polyurethane or marine varnish on the exterior wood.

5. Stack all the pieces together (see the following figure).

Pick the spot where you want to locate your hive and place the elevated hive stand on the ground. The stand provides the bees with good ventilation, keeps the hive off the damp ground, and raises the colony so it's much easier for you to inspect.

Place the hive body centered on top of the stand. The end of the hive with the holes is the entrance, so be sure the hive is facing in the direction you intend for the bees to fly. See Chapter 2 for ideas on where to locate your hive.

Fill the hive body with the top bars, all 28 of them. The bees will build their beautiful comb on the underside of each top bar along the waxed starter strips. Here they'll raise their brood and store pollen and honey.

Now top everything off with the ventilated roof. It provides your bees with protection from the elements. That's it! You're ready for the bees!

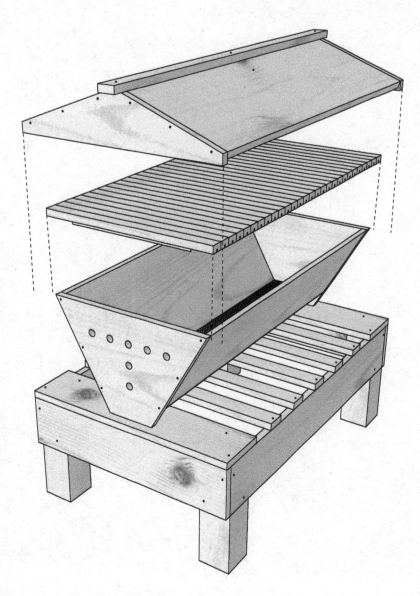

Illustration by Felix Freudzon, Freudzon Design

Chapter 6
The Five-Frame Nuc Hive

Illustration by Felix Freudzon, Freudzon Design

A nuc hive (sometimes called a *nucleus hive* or simply a *nuc*) is a small hive with a few frames of bees. The design is very similar to the Langstroth hive that I cover in Chapter 10, only with fewer frames. Why create this smaller capacity hive? Here are some reasons:

- A nuc can serve as a nursery for raising new queens.

- A nuc provides you with a handy source of brood, pollen, and nectar to supplement weaker colonies (kind of like having your own dispensary).

- You can sell a nuc colony of bees to other beekeepers — they're a fast way to populate a new hive.

- You can use a nuc to populate an observation hive.

✔ You can use a nuc to house a captured swarm of bees.

✔ You can use a nuc to house a supply of bees used for bee-venom sting therapy. You can find more information on the use of bee venom to treat certain inflammatory medical conditions from the American Apitherapy Society (www.apitherapy.org).

✔ A nuc in the corner of a garden can help with pollination and requires far less maintenance than a regular hive (but you won't harvest any honey from this little hive).

The one disadvantage of a nuc hive is that it doesn't overwinter well in colder climate zones because it doesn't have enough bees or stored honey to see the little colony through the winter months. So if you live in an area where the winters are cold (long periods below freezing), you should combine your nuc colony with one of your big hives before Jack Frost pays a visit. However, if you have mild winters (few or no days under about 40 degrees Fahrenheit), you can feed the nuc colony using an entrance feeder or a baggie feeder and it should do fine. (An *entrance feeder* consists of a small inverted jar of syrup that sits in a contraption at the entrance to a hive.)

Vital Stats

✔ **Size:** 23 inches x 11 inches x 13¼ inches.

✔ **Capacity:** Because this design consists of only five frames, there's no room for expansion as the colony grows in population, so the capacity for bees is limited. Theoretically, you could build additional nuc hive bodies and stack them one on top of the other to allow the colony to grow. But that's not really the purpose of a nuc hive, unless you're using the nuc as a pollination source in your garden.

✔ **Type of frame:** This nuc hive uses a Langstroth-style, self-centering frame with beeswax foundation inserts. It has a total of five deep frames.

✔ **Universality:** Because the nuc uses Langstroth-style frames, you can easily purchase them for this hive (they're available from beekeeping supply stores). If you want to try building your own frames for this hive, check out Chapter 17.

✔ **Degree of difficulty:** This is a pretty straightforward design. However, the metalwork involved with the aluminum flashing used on the outer cover can be a challenge (bending the corners takes a little patience and practice).

✔ **Cost:** Using scrap wood (if you can find some) would keep material costs of this design minimal. But even if you purchase all the recommended lumber, hardware, and fasteners, you can likely build this nuc hive for about $85 (a little less if you use knotty pine lumber).

Materials List

The following table lists what you'll use to build your nuc hive. In most cases, you can make substitutions as needed or desired. (See Chapter 3 for help choosing alternate materials.)

Lumber	Hardware	Fasteners
1 8' length of 1" x12" clear pine lumber	1 roll of 14"-wide aluminum flashing (usually comes in a 10' length)	80 #6 x 1⅜" deck screws, galvanized, #2 Phillips drive, flat-head with coarse thread and sharp point
1 2' x 4' sheet of ¾" thick exterior plywood	Optional: weatherproof wood glue	25 #8 x ½" lath screws, galvanized, #2 Phillips drive, flat-head with sharp point
1 2' x 4' sheet of ¼" thick lauan plywood	Optional: A quart of latex or oil exterior paint (white or any light color), exterior polyurethane, or marine varnish	

Here are a few tips and tricks to purchasing materials for your five-frame nuc hive:

✔ I'm a big fan of clear pine. It's not too expensive as lumber goes. Alternatively, knotty pine is even less expensive. You can also use different kinds of wood for your nuc hive. Cedar and cypress make beautiful hives, and you can really get fancy with a cherry or mahogany nuc hive. It's up to you.

✔ Depending on where you buy it, plywood sometimes comes as ²³⁄₃₂ inch (rather than ¾ inch). No worries: The difference is minimal, and either way, the plywood will fit just fine.

✔ In the case of fasteners, I've tossed in a few more pieces than you'll use because, if you're like me, you'll lose or bend a few along the way. It's better to have a few extras on hand and save another trip to the hardware store.

Cut List

The following sections break down the five-frame nuc hive into its individual components and provide instructions on how to cut and build those components.

Lumber in a store is identified by its *nominal* size, which is its rough dimension before it's trimmed and sanded to its finished size at the lumber mill. The actual finished dimensions are always slightly different from the nominal dimensions. For example, what a lumberyard calls *1 inch x 12 inch lumber* is in fact ¾ inch x 11¾ inch. In the following sections, each Material column lists nominal dimensions, and each Dimensions column lists the actual, final measurements.

This hive makes use of dado and rabbet joints. See Chapter 4 for detailed instructions on making all these cuts.

Bottom board

Quantity	Material	Dimensions	Notes
2	1" x 12" clear pine	22" x 1⅛" x ¾"	These are the side rails. Dado ¾" wide x ⅜" deep along the entire length of each side rail, and rabbet one of the rear corners of each opposite side rail ¾" wide x ⅜" deep.
1	1" x 12" clear pine	21⅝" x 8¼" x ¾"	This is the "floor."
1	1" x 12" clear pine	8¼" x 1⅛" x ¾"	This is the rear rail. Dado ¾" wide x ⅜" deep along the entire length (see detail for placement of dado cut).
1	1" x 12" clear pine	7⅜" x ¾" x ¾"	This is the entrance reducer. Cut two notches on two different sides of the entrance reducer as shown in the figure (one side ⅜" high x ¾"wide, and the other side ⅜" high x 4" wide).

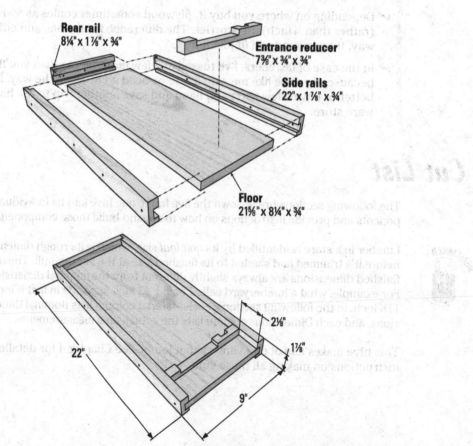

Rear rail
8¼" x 1⅛" x ¾"

Entrance reducer
7⅜" x ¾" x ¾"

Side rails
22" x 1⅛" x ¾"

Floor
21⅝" x 8¼" x ¾"

2⅛"

1⅛"

22"

9"

Illustration by Felix Freudzon, Freudzon Design

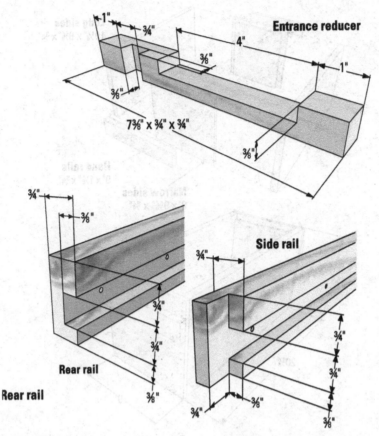

Entrance reducer

1"
¾"
4"
⅜"
1"
⅜"
7⅜" x ¾" x ¾"
⅜"

¾"
⅜"
Side rail
¾"
¾"
¾"
¾"
Rear rail
¾"
Rear rail
⅜"
¾"
⅜"
⅜"

Illustration by Felix Freudzon, Freudzon Design

Hive body

Quantity	Material	Dimensions	Notes
2	1" x 12" clear pine	19⅛" x 9⅝" x ¾"	These are the long sides.
2	1" x 12" clear pine	9" x 9⅝" x ¾"	These are the narrow sides. Rabbet ⅝" wide x ⅜" deep along the entire inside top length of both panels; rabbet a ¾" wide x ⅜" deep joinery cut along both ends of the narrow panels.
2	1" x 12" clear pine	9" x 1⅞" x ¾"	These are the hand rails.

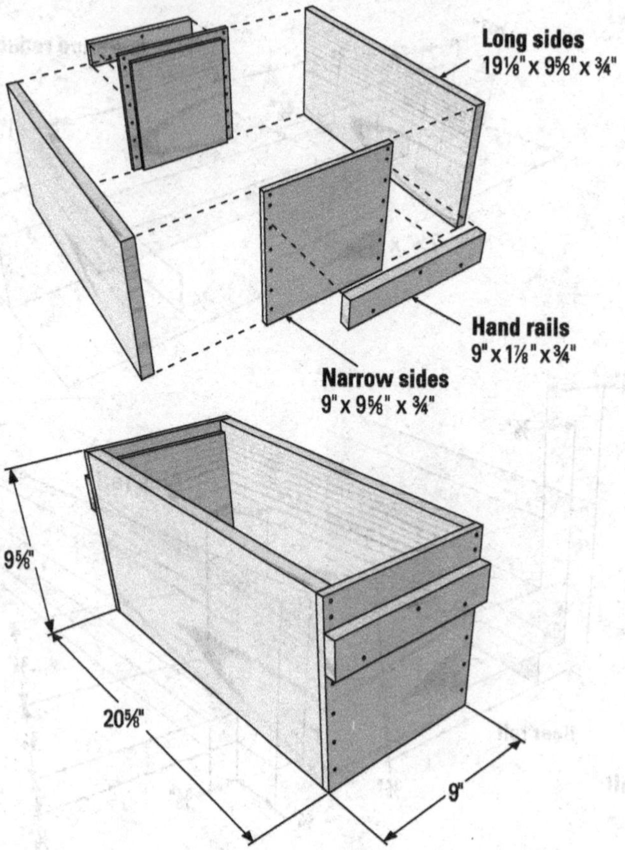
Long sides
19⅛" x 9⅝" x ¾"
Hand rails
9" x 1⅞" x ¾"
Narrow sides
9" x 9⅝" x ¾"
9⅝"
20⅝"
9"

Illustration by Felix Freudzon, Freudzon Design

Inner hive cover

Quantity	Material	Dimensions	Notes
2	1" x 12" clear pine	19⅞" x ¾" x ¾"	These are the long rails. Dado ¼" wide x ⅜" deep along entire length, ⅛" from edge.
2	1" x 12" clear pine	8¼" x ¾" x ¾"	These are the short rails. Dado ¼" wide x ⅜" deep along entire length, ⅛" from edge. Optional: Cut a ¾" wide x ¼" deep ventilation notch at center point of one short rail (on the thick side of the rail).
1	¼" lauan plywood	19⅛" x 8½" x ¼"	This is the top. Drill a 1" round ventilation hole in the center of the plywood.

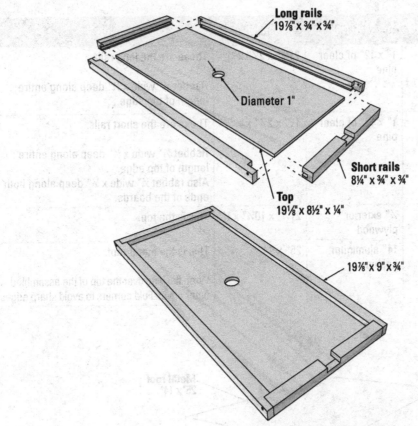

Long rails
19⅞" x ¾" x ¾"

Diameter 1"

Short rails
8¼" x ¾" x ¾"

Top
19⅛" x 8½" x ¼"

19⅞" x 9" x ¾"

Illustration by Felix Freudzon, Freudzon Design

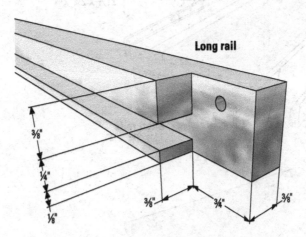

Long rail

⅜"

¼"

⅛"

⅜"   ¾"   ⅜"

Illustration by Felix Freudzon, Freudzon Design

Outer hive cover

Quantity	Material	Dimensions	Notes
2	1" x 12" of clear pine	21¼" x 2¼" x ¾"	These are the long rails. Rabbet ¾" wide x ⅜" deep along entire length of top edge.
2	1" x 12" of clear pine	11" x 2¼" x ¾"	These are the short rails. Rabbet ¾" wide x ⅜" deep along entire length of top edge. Also rabbet ¾" wide x ⅜" deep along both ends of the boards.
1	¾" exterior plywood	21¼" x 10¼" x ¾"	This is the top.
1	14" aluminum flashing	25" x 14"	This is the metal roof. Wrap flashing over the top of the assembled outer cover. Fold corners to avoid sharp edges.

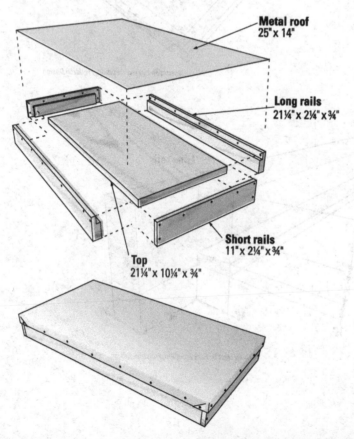

Metal roof
25" x 14"

Long rails
21¼" x 2¼" x ¾"

Short rails
11" x 2¼" x ¾"

Top
21¼" x 10¼" x ¾"

Illustration by Felix Freudzon, Freudzon Design

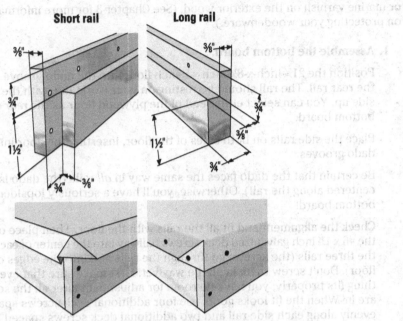

Short rail   Long rail

Illustration by Felix Freudzon, Freudzon Design

Assembling the Hive

Now that you have all the pieces cut, it's time to put it all together and build your nuc. Start at the bottom (the ground) and work your way up (the sky).

When screwing together parts in the following steps, remember that the screws will go in easier if you first drill a ⁷⁄₆₄-inch hole in each spot you plan to place a screw. The pre-drilling also helps prevent the wood from splitting.

Consider using a weatherproof wood glue in addition to the deck screws. It helps make the components as strong as possible. Apply a thin coat of glue wherever wooden parts are joined together.

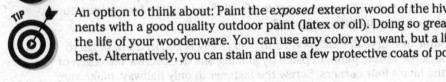

An option to think about: Paint the *exposed* exterior wood of the hive's components with a good quality outdoor paint (latex or oil). Doing so greatly extends the life of your woodenware. You can use any color you want, but a light color is best. Alternatively, you can stain and use a few protective coats of polyurethane or marine varnish on the exterior wood. (See Chapter 3 for more information on protecting your woodenware.)

1. **Assemble the bottom board.**

   Position the 21⅝-inch-x-8¼-inch-x-¾-inch floor into the dado groove of the rear rail. The rail should be resting on your worktable with the dado side up. You can select either end of the plywood floor as the *rear* of the bottom board.

   Place the side rails on both sides of the floor, inserting the floor into the dado grooves.

   Be certain that the dado faces the same way in *all* rails (the dado isn't centered along the rail). Otherwise, you'll have a seriously lopsided bottom board!

   Check the alignment and fit all the rails with the floor. Then place one of the #6 x 1⅝ inch galvanized deck screws halfway into the center of each of the three rails (the screws go through the rails and into the edges of the floor). Don't screw them in all the way yet. First make sure that everything fits properly; you have no room for adjustment after all the screws are in! When the fit looks good, use four additional deck screws spaced evenly along each side rail and two additional deck screws spaced evenly along the rear rail.

   The entrance reducer remains loose, and you place it in the hive's entrance to control ventilation and prevent *robbing,* which is when bees from other colonies invade your hive and steal your colony's precious honey. The entrance reducer is typically *not* used year-round. For more information on using an entrance reducer, see Chapter 2.

2. **Assemble the hive body.**

   On the narrow rabbeted sides of the hive body, use a ⁷⁄₆₄-inch bit to drill guide holes where you plan to place screws (these holes make it easier for the screws to go in and prevent the wood from splitting). Drill seven holes per rabbeted corner (see the following figure for approximate spacing).

   Assemble the two long sides and the two narrow sides of the hive body by placing the joints together. You're essentially building a box.

   Use a carpenter's square to make sure the box stays square as you assemble the hive body because you won't have an opportunity for correction after all the screws are in place! (See Chapter 4 for more information on keeping things square.)

   After the "dry" fit looks good and everything is squared up, begin to lock the rabbet joints in place by affixing one deck screw into each of the hive's four corners. Screw the fastener in only halfway, make sure everything remains square and fits properly, and then screw this and the remaining fasteners all the way in. Use a total of seven deck screws per rabbeted corner, spaced approximately as indicated in the following figure.

Now use deck screws to attach the two hand rails to the narrow sides of the nuc hive body. Position the top edge of the hand rails 2 inches down from the top edge of the nuc body. Use three screws per hand rail, spaced and staggered (to prevent splitting the wood). Precise placement of screws isn't critical.

In place of using wooden hand rails, you can attach flush-mounted, galvanized (or stainless steel) handles to the nuc hive body. They look nice, and they give you a much better grip when lifting the hive body. You can find these handles in hardware stores or marine supply stores.

Check all sides to make certain that all the screws are in place.

3. **Assemble the inner cover.**

   Position the plywood cover insert into the dado grooves of the long rails and the short rails. It's kind of like putting a picture frame together.

   Be certain that *all* rails have the *thick* or *thin* lip of the groove facing the same way. Otherwise, you'll have a seriously lopsided inner cover!

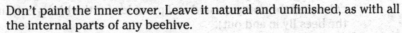

   Check the alignment and fit, and insert a deck screw halfway into each of the four corners. Make sure everything is square and fits properly and then screw them in all the way. Keep in mind that if you cut the plywood insert perfectly square, it will square up the frame.

   Don't paint the inner cover. Leave it natural and unfinished, as with all the internal parts of any beehive.

   Note: You should position the inner cover on the hive body with the flat side down and with the cutout notch (bee ventilation/entrance) facing up and at the front of the hive.

4. **Assemble the outer cover.**

   Start with one long rail. Insert the plywood onto the shelf created by the rabbeted groove. Repeat this step on the opposite side with the second long rail.

   Fit both of the two short rails onto the plywood board. The short and long rails form a frame surrounding the plywood board. If the plywood was cut perfectly square, it will square up the entire assembly.

   When assembling the outer cover, it's helpful to have a "stop" on your worktable that you can push against while inserting screws. A short piece of 2x4 lumber clamped or screwed to the table serves as a good stop to work against.

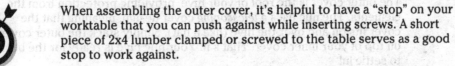

   Place one end of the outer cover flat on the worktable against the stop. Insert two deck screws into each corner of the short rails. Reverse the entire cover end to end and screw the other corners of the short rails in a similar manner. Make certain the entire assembly remains snug and tight as you do this.

Now use additional deck screws to secure the plywood insert to the frame assembly. Drive the screws through the rails and into the edges of the plywood board. Five evenly spaced screws along each long rail and three along the short rails should do the trick (see the figure in the earlier section "Outer hive cover").

Center the aluminum flashing evenly on the top of the outer cover, and bend the flashing over the edges of the rail/frame. This creates a lip all around the top edge. Do this to all four sides. Bend and fold the corners (as if you're making the corners of a bed). The flashing is thin and fairly easy to work with. Use a rubber mallet to coax the corners flush and flat.

The edges of aluminum flashing are very sharp. Use caution when handling flashing to avoid cutting yourself, and consider using work gloves.

Affix the flashing's folded edges to the outer cover using the #8 x ½ inch lath screws. I use six evenly spaced lath screws per long side (plus an extra screw to secure each folded corner) and five screws per short side.

5. **Stack all the pieces together to create the nuc hive.**

   Now it's time to put all the elements together. Place the bottom board on level ground. The bottom board is the nuc hive's floor. It keeps the colony off the damp ground and provides for the hive's entrance (where the bees fly in and out).

Consider using an elevated hive stand to raise the nuc farther off the ground and make it more accessible to you. (See Chapter 13 for instructions on building an elevated hive stand.)

The nuc body goes on top of the bottom board. In this box, the bees raise baby bees and store food for their use. Place five deep frames with foundation (either store-bought or ones you make yourself) into the nuc body.

Place the inner cover on top of the hive body. The deeper ledge faces up. The ventilation notch in the inner cover faces upward and toward the front (entrance) of the nuc hive.

The outer cover is the roof of your hive, providing protection from the elements. The nuc uses a _telescoping_ cover design, meaning that the cover fits on and over the hive (like a hat). Simply stack the outer cover on top of your inner cover. That's it. Your nuc hive is ready for the bees to settle in!

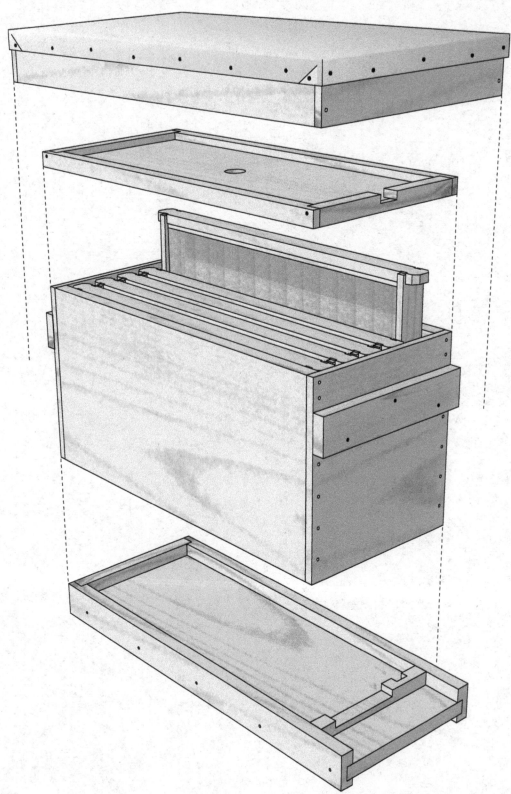

Illustration by Felix Freudzon, Freudzon Design

Chapter 7
The Four-Frame Observation Hive

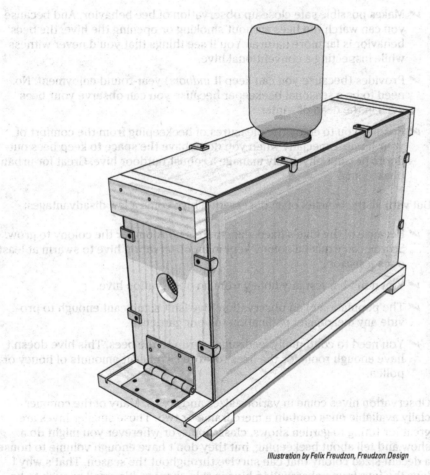

Illustration by Felix Freudzon, Freudzon Design

An *observation hive* is a small hive with glass panels that enables you to observe a colony of bees without disturbing them or risking being stung. Such hives are usually kept indoors but typically provide access for the bees to fly freely from the hive to the outdoors (the hive in this chapter has an entry that you can open and close).

I'm a big believer in having an observation hive, even when you have conventional hives in your garden. The pleasure and added insight it gives you about honeybee behavior is immeasurable. A few of the rewards you can realize from setting up an observation hive are that it

✔ Serves as a fantastic educational tool to take to schools, farmers markets, garden shows, and so on.

✔ Gives you a barometer on what's happening in a bee colony at any given time of year. That way you can anticipate the needs of your outdoor colonies and better manage your hives.

✔ Makes possible safe close-up observation of bee behavior. And because you can watch the bees without smoking or opening the hive, the bees' behavior is far more natural. You'll see things that you'd never witness while inspecting a conventional hive.

✔ Provides (because you can keep it *indoors*) year-round enjoyment. No need to be a seasonal beekeeper because you can observe your bees even in the dead of winter.

✔ Enables you to enjoy the pleasures of beekeeping from the comfort of your home, especially when you don't have the space to keep bees outdoors or can't physically manage a robust outdoor hive. Great for urban beekeeping!

But with all the benefits of an observation hive come a few disadvantages:

✔ Because of the hive's fixed size, there's no room for the colony to grow. So you can expect a colony kept in an observation hive to swarm at least once a season.

✔ You don't harvest any honey from an observation hive.

✔ The population of an observation hive isn't significant enough to provide any meaningful pollination to your garden.

✔ You need to continually feed sugar syrup to the bees. This hive doesn't have enough room for the bees to store significant amounts of honey or pollen.

Observation hives come in various sizes and styles. Many of the commercially available ones contain a mere frame or two. These smaller hives are great for toting to garden shows, classrooms, or wherever you might do a show and tell about beekeeping, but they don't have enough volume to house a decent-sized colony that can survive throughout the season. That's why I love the four-frame observation hive in this chapter. It's small enough to be portable for educational demonstrations but substantive enough (with four deep frames) to fare well throughout the season (even throughout the winter months).

Keeping bees in an observation hive requires a different skill set than raising bees in a typical hive. For detailed information, I recommend the book *Observation Hives: How to Set Up, Maintain and Use a Window to the World of Honey Bees,* by Thomas Webster and Dewey Caron (A. I. Root Company).

Vital Stats

- ✔ **Overall size:** 22 inches x 7¼ inches x 12⅜ inches.

- ✔ **Capacity:** Because this design consists of only four frames and allows no room for expansion as the colony grows in population, the capacity for bees is limited. So prepare yourself for the inevitable — bees kept in an observation hive will swarm at some point.

- ✔ **Type of frame:** This hive uses a Langstroth-style, self-centering frame with beeswax foundation inserts (flip to Chapter 10 for an introduction to the Langstroth hive). The hive uses a total of four deep frames.

- ✔ **Universality:** Because this hive uses Langstroth-style deep frames, you can easily purchase ready-made deep frames and deep foundation from any beekeeping supplier. (If you really want to make your own deep frames, check out Chapter 17.)

- ✔ **Degree of difficulty:** This is a simple design that doesn't have too many parts and is easy to build.

- ✔ **Cost:** Using scrap wood (if you can find some) would keep material costs of this design minimal, but even if you purchase the recommended lumber, hardware, glass, and fasteners, you can likely build this observation hive for less than $40. The most expensive items are the two panes of tempered window glass. If you choose Plexiglas, be prepared to spend more (it's pricey).

Materials List

The following table lists what you'll use to build your observation hive. In most cases, you can make substitutions in lumber as needed or desired. (See Chapter 3 for help choosing alternate materials.)

Lumber	Hardware	Fasteners
1 8' length of 1" x 8" knotty pine lumber	10" x 10" piece of ⅛" hardware cloth (#8 hardware cloth is usually sold in 3' x 10' rolls, but some beekeeping suppliers offer it by the foot)	20 ⅜" staples for use in a heavy-duty staple gun
	2" x 4" wide throw door hinge	14 #6 x 1⅝" deck screws, galvanized, #2 Phillips drive, flat-head with coarse thread and sharp point
	4 Deep Langstroth-style frames and deep foundation (available from any beekeeping supply source)	20 ⅛" flat-head Phillips screws
	2 ⅛" panes tempered window glass cut to 19¾" x 11" (have glazer round off all sharp edges)	6 ½" flat-head Phillips screws
	Mason jar or empty mayonnaise jar with metal screw-on lid that measures slightly less than 3" in diameter	12 ⅝" x ⅜" bendable "L" mirror clips and screws (clips typically come with matching screws)
	Optional: weatherproof wood glue	
	Optional: wood stain and polyurethane finish	

Here are a few notes about the materials for your four-frame observation hive:

✔ I specify using knotty pine because it's about the least expensive wood on the market. However, because this hive will be on display in your home or at educational events, you may want to invest in some fancy lumber to dress it up for show! Consider a stunning cherry wood hive or even a jaw-dropping veneer hive (Chapter 19 shows my fancy version of this hive).

✔ Using *tempered* (safety) glass (versus normal glass) greatly reduces the chance of breakage and injury. Alternatively, you can use Plexiglas PMMA resin, which results in a virtually unbreakable window. However, over time, Plexiglas scratches and gets cloudy, and it's much more difficult to clean than glass. I prefer tempered glass.

✔ I've tossed in a few more screws and nails than you'll actually use because, if you're like me, you'll lose or bend a few along the way. It's better to have a few extras on hand and save yourself another trip to the hardware store.

Cut List

The following sections break down the observation hive into its individual components and provide instructions on how to cut those components.

Lumber in a store is identified by its *nominal* size, which is its rough dimension before it's trimmed and sanded to its finished size at the lumber mill. The actual finished dimensions are always slightly different from the nominal dimensions. For example, what a lumberyard calls *1 inch x 8 inch lumber* is in fact ¾ inch by 7½ inch. In the following tables, each Material column lists nominal dimensions, and each Dimensions column lists the actual, final measurements.

This hive requires dado and saw kerf cuts. See Chapter 4 for detailed instructions on making these cuts.

Hive body and top

Quantity	Material	Dimensions	Notes
2	1" x 8" knotty pine	10⅞" x 5¾" x ¾"	These are the side panels. Dado a ¾" wide by ⅜" deep channel along the entire inside top width of both side panels. Make the dado cut ¾" down from the top edge (the four frames rest on the ledge this dado creates). Drill an entrance hole 1½" in diameter, centered left to right, and positioned ¼" from the bottom edge of *one* of the side panels. This is the entrance for the bees. Drill a 1½" ventilation hole centered left to right and top to bottom on *both* side panels.
2	1" x 8" knotty pine	5¾" x 1¾" x ¾"	These are the hive's handles.
1	1" x 8" knotty pine	18¾" x 5¾" x ¾"	This is the top panel (roof) of the hive. Drill a feeding hole 3" in diameter through the top, centered on the board. This hole accommodates the lid of the feeding jar.
3	#8 hardware cloth	4" x 4"	This screening is attached to the top panel (roof) and to the ventilation holes in the side panels.

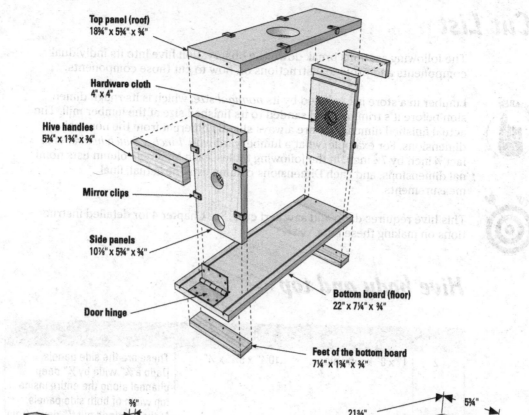

Top panel (roof)
18¾" x 5¾" x ¾"
Hardware cloth
4" x 4"
Hive handles
5¾" x 1¾" x ¾"
Mirror clips
Side panels
10⅞" x 5¾" x ¾"
Door hinge
Bottom board (floor)
22" x 7¼" x ¾"
Feet of the bottom board
7¼" x 1¾" x ¾"
⅜"
¾"
¾"
4 9/16"
Diameter 1½"
10⅞"
1⅛"
Diameter 1½"
¾"
21⅜"
5¾"
Diameter 3"
10⅞"
19⅞"
22"
7¼"

Illustration by Felix Freudzon, Freudzon Design

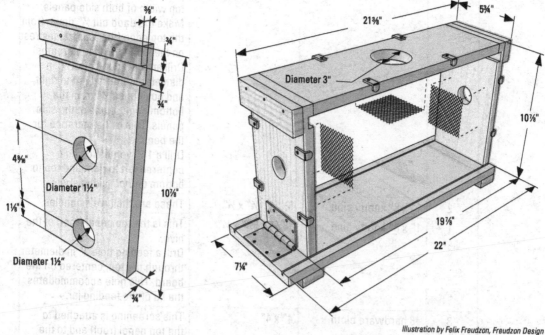

Bottom board

Quantity	Material	Dimensions	Notes
2	1" x 8" knotty pine	7¼" x 1¾" x ¾"	These are the "feet" of the bottom board.
1	1" x 8" knotty pine	22" x 7¼" x ¾"	This is the bottom (floor) of the hive. Cut saw kerf grooves, ⅛" wide and ¼" deep, along the entire long length of the bottom board, centered and placed 5¾" apart. These grooves hold the bottom of the glass panels in place. Check the fit of the glass panels in the grooves before assembling the hive; adjust the kerf cut if needed to accommodate the thickness of the glass. The fit should be snug.

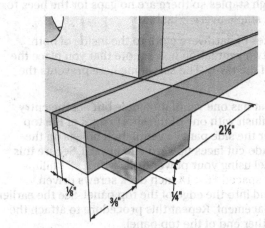

2⅛"

⅛"

⅜"

¼"

Illustration by Felix Freudzon, Freudzon Design

Assembling the Hive

Now that you've cut all the parts for your four-frame observation hive, you're ready to put it all together.

Here are a few pointers to follow throughout this assembly:

- Consider using a weatherproof wood glue in addition to screws. It helps make the observation hive as strong as possible. Apply a thin coat of glue wherever the wooden parts are joined together.

- Screws will go in easier if you first drill a 7/64 inch hole in each spot you plan to place a screw. The pre-drilling also helps prevent the wood from splitting.

- Use a carpenter's square to make sure everything stays square as you put the hive together; you have no opportunity for correction after all the screws are in place! (See Chapter 4 for information on keeping things square.)

You start at the top and work your way down:

1. **Attach the two side panels of the hive body to the top panel.**

   Staple one piece of hardware cloth to cover the feeding hole (note that you place the screening on the *inside* of the top panel). The screening here prevents the bees from escaping when you remove the feeding jar for refilling. Use enough staples so there are no gaps for the bees to escape through. I place a staple every ½ inch.

   Staple the two other pieces of hardware cloth to the inside of both side panels to cover the two ventilation holes (note that you place the screening on the *inside* of the hive). The screening here prevents the bees from escaping.

   Identify the side panel that has one ventilation hole but *not* the entry hole. Position this panel flush with one of the short edges of the top panel. The screening over the side panel's ventilation hole is on the inside of the hive. The dado cut faces upward and inward. Secure this side panel to the top panel using your power drill (with a #2 Phillips head bit) and two evenly spaced #6 x 1⅜ inch deck screws driven through the side panel and into the edge of the top panel. See the earlier figure for approximate placement. Repeat this procedure to attach the other side panel to the other end of the top panel.

2. **Attach the hand rails to the side panels.**

   Use deck screws to attach the hand rails to the side panels. Note that the top edge of the hand rails are flush with the top. The screws go through the hand rails and into the side panels. Use three evenly spaced screws per hand rail. Be certain to position the screws to avoid the screws you've already used to attach the side panels to the top panel. See the following figure for approximate placement of screws.

3. **Attach the top assembly to the bottom board.**

You center the entire top assembly between the two ⅛ inch kerf cuts on the bottom board. Position the panel that doesn't have the entrance hole flush with one short edge of the bottom board. Use two evenly spaced deck screws per side, driven through the underside of the bottom board and into the bottom edge of the side panels. See the earlier figure for approximate placement of screws.

4. **Attach the feet to the bottom board.**

Take one of the feet and position it under the bottom board, centered below one of the side panels. Secure the foot to the bottom board using three evenly spaced deck screws. The screws go through the foot and into the bottom board. Repeat this procedure to attach the second foot. Be certain to position the screws to avoid the screws you've already used to attach the top assembly to the bottom board.

Optional: Stain the *exposed* wood of the observation hive and use a few protective coats of polyurethane or marine varnish. (See Chapter 3 for information on protecting your woodenware.)

5. **Attach the hinge hardware to the entry area.**

You need a door that you can open and close to let the bees fly out or to keep them inside. The door feature should be foolproof, so that when you take your observation hive to schools, curious little students can't open it and let all the bees out. (That would make for a memorable school day!) A wide door hinge serves perfectly as a door (in this case, a 2-inch by 4-inch wide-throw door hinge). Secure one flap of the hinge to the bottom board using the ½ inch flat-head Phillips screws, and position the other flap to close the entry hole when the flap is up. When the flap is down, the entrance hole is open.

For closing the entrance, use a screw or two to secure the hinge flap in the up position. That prevents the curious from flipping the door open.

6. **Install four deep frames with foundation.**

The frames slide into place and hang on the "ledges" of the side panels that you created with the dado cuts.

7. **Attach the glass window panels.**

Insert one of the tempered glass panels into the ⅛ inch kerf cut that runs along the length of the bottom board. After you center the glass panel on one side of the hive, secure the glass snugly in place by using six evenly spaced "L" mirror clips (two along each side and two along the top). Use one ⅛ inch flat-head Phillips screw per clip, or use the screws that typically come with mirror clips.

TIP

Some mirror clips come with a patch of felt attached to the surface of the clip that comes in contact with the glass. If your clips don't have this feature, cut a small square of felt and glue it to the facing of the clip that pairs with the glass. This holds the glass more snugly, prevents the glass from scratching, and eliminates any rattles.

Repeat this procedure for the glass window panel on the opposite side.

8. Top off your hive with the feeding jar.

Use a brad to punch a dozen or so tiny holes in the metal lid of a Mason jar or empty mayonnaise jar; these holes allow the bees access to the syrup. Fill the feeding jar with sugar syrup and invert it over the hole in the top of the hive. That's it! Your hive is ready for a colony of bees and many hours of enjoyable observation.

REMEMBER

To install bees, you need to remove one of the glass window panels to access the frames. The easiest way to install bees in an observation hive is to replace one or more of the new empty frames with an equal number of frames of drawn comb containing capped brood, worker bees, and a queen.

TIP

A simple recipe for syrup is to dissolve one pound of granulated sugar in one pint of very hot water. Let the syrup cool before feeding it to the bees. Keep unused syrup refrigerated until you're ready to use it.

REMEMBER

When *not* using your observation hive for show and tell, you need to allow the bees free access in and out of the hive. Never leave them trapped inside for more than 24 hours. By flying in and out, your girls can gather food and have their "cleansing flights" (go to the toilet). If you keep your hive *outside* during nice weather, make sure the hinge covering the access hole is left open so the bees can travel in and out. Keep the hive out of direct sun and under shelter from the rain, and only keep it outside when the temperature is above 60 degrees.

If you keep your bees inside (as you'd need to do anyway during cold weather), they still need to be able to fly in and out. In that case, you can purchase a length of 1½ inch diameter clear plastic or vinyl tubing (available from most hardware stores) and insert it snugly into the entrance hole. The other end of the tubing leads to the outside of the house, through either a 1½ inch diameter hole drilled through a wall (too destructive for me) or a 1½ inch diameter hole drilled through a rectangular piece of plywood. You can place the plywood below a partially opened double hung window. Cut the plywood to the width of the window casing. It only needs to be a few inches tall.

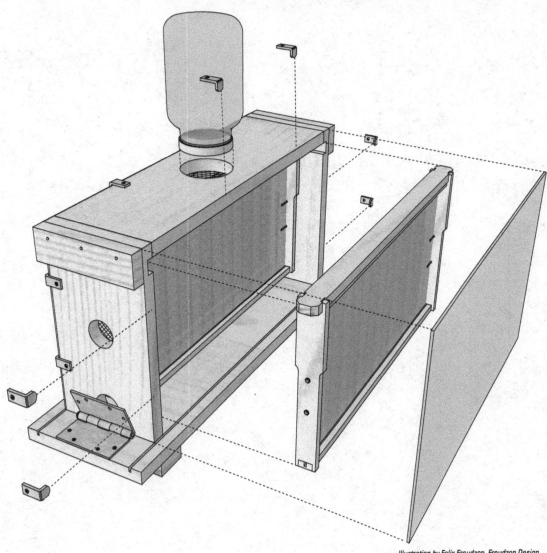

Illustration by Felix Freudzon, Freudzon Design

Chapter 8
The Warré Hive

Illustration by Felix Freudzon, Freudzon Design

The Warré hive was developed in France during the early 20th century by Abbé Émile Warré, an ordained priest and avid beekeeper. His vision was to develop an easy-to-build and easy-to-manage beehive that everyone could have success with (thus often referred to as the *people's hive*). This simple, cost-effective, and efficient design has been gaining renewed popularity among DIY beekeepers and those seeking more natural approaches to bee-keeping. I'm enthusiastic about this design and appreciate how it provides the bees with living conditions more like those found in nature.

Here are some of the pros of the Warré hive:

- The top bars (which have no bottom rails or side rails and use no full sheets of foundation) allow the bees to build their comb freestyle and without restrictions on cell size.

- The vertical design provides a living arrangement that's similar to how bees live in the wild (working downward from top to bottom and within the hollow of a tree). As a result, the bees tend to be less stressed and thus less prone to disease.

- As building beehives go, the Warré hive is simple to build (no tricky woodworking and no frame building) and quite cost-effective.

- As per Warré management protocol, this design demands very little interaction on the beekeeper's part. It's kind of a "set it and forget it" style of beekeeping, with nearly no inspections, other than adding supers and harvesting honey at the end of the season (inspecting a Warré hive often damages the delicate, natural comb). Less inspection means less heat loss, better natural control of temperature and humidity, and less stress on the bees.

However, with the good comes some bad. Here are a few negatives associated with this design:

- To harvest the honey you must destroy the comb (you have to crush it to extract the honey). On the other hand, every year, the colony uses a fresh supply of beeswax comb, so potential pesticide buildup is reduced.

- The honey you harvest will contain a high percentage of pollen, making it cloudy and "less pretty" if you plan to sell it. But this is actually a good thing. Such honey is healthier than the supermarket stuff — that extra pollen means more protein.

- Doing various beekeeping manipulations is difficult or even impossible versus what's possible with a hive that uses framed comb. The comb in a Warré hive is far too fragile to be moving around. But this *negative* is actually a *positive*. The idea of the Warré is to provide as natural an environment as possible for your bees. Manipulations are contrary to the all-natural objective.

- Administering medications can be difficult (the hive has no feeder, for example). But the Warré hive is all about sustainability and a more natural approach to beekeeping. Medications and chemicals need not apply. Again, with this negative lives a positive. The bees tend to be healthier when under less stress and in a more natural environment.

Vital Stats

- **Overall size:** 18 inches x 19¾ inches x 47⁷⁄₁₆ inches.

- **Capacity:** The tall, vertical design, made up mostly of four hive boxes, provides ample room for the colony to naturally grow throughout the season.

- **Type of frame:** This design uses a top-bar style of frame. It has no side bars or bottom bar and no full sheets of beeswax foundation to deal with (or the costs associated with these elements). The bees build their comb naturally and without restriction onto each of the top bars in the hive (32 frames in all).

✔ **Universality:** This characteristic is less important with this design than with others. All the gadgets and add-ons that you might use with a conventional Langstroth hive (feeders, queen excluders, foundation, and honey-extracting equipment) are irrelevant to the Warré. Its design is virtually all-inclusive and requires no extras. However, if you need a new roof and don't want to make another one, an increasing number of beekeeping supply stores offer prebuilt Warré components. Just make sure the dimensions of the commercially made hives jibe with your dimensions (Warré measurements are less standardized than those of the more-popular Langstroth hive, which I describe in Chapter 10).

✔ **Degree of difficulty:** As beehives go, this is likely one of the easiest to build. It has no complex joinery, and you don't have to build complicated frames or insert foundation. All in all, this design has fewer components to deal with than a Langstroth or British National hive. It's a winner!

✔ **Cost:** Using scrap wood (if you can find some) would keep material costs of this design minimal, but even if you purchase the recommended knotty pine, hardware, and fasteners, you can likely build this Warré hive (top bar frames and all) for less than $75.

Materials List

The following table lists what you'll use to build your Warré hive and the top bar frames used with it. In most cases, you can make substitutions as needed or desired. (See Chapter 3 for help choosing alternate materials.)

Lumber	Hardware	Fasteners
5 8' lengths of 1" x 10" knotty pine lumber	½ pound beeswax (to melt and brush on the frames' starter strips)	10 #6 x ⅝" wood screws, #2 Phillips drive, flat-head
1 8' length of 2" x 3" knotty pine	Burlap (hessian) cloth, available at hardware and garden supply stores; you only need a small piece (13⅝16" x 13⅝16"). You may need to purchase a roll of burlap and cut to this dimension.	70 #6 x 1⅝" deck screws, galvanized, #2 Phillips drive, flat-head with coarse thread and sharp point
1 2' x 4' sheet of ⅜" thick exterior plywood	Enough insulation material to fill the quilt box — you can use dry leaves, straw, wood chips, pine needles, or coarse sawdust	100 6d x 2" galvanized nails
1 12" x 36" x ³⁄₃₂" balsa wood sheet	Optional: weatherproof wood glue	20 ⅜" staples for use in a heavy-duty staple gun
	Optional: a quart of latex or oil exterior paint (white or any light color), exterior polyurethane, or marine varnish	

Here are a couple of notes about the materials for your Warré hive:

✔ Because of its simple butt-joinery, knotty pine is perfectly okay for this hive, and it's about the least expensive board lumber out there. But pine isn't the most durable wood, so it benefits from a protective coating of paint, varnish, or polyurethane. Alternatively, consider keeping your hive as all-natural as possible by using cedar or cypress. Although these woods are more expensive than knotty pine, they stand up well to weather without paint or protective chemicals.

✔ You can purchase ³⁄₃₂ inch thick sheets of balsa wood online or from a hobby store. Use a utility knife to cut the balsa sheets into the 10 inch x ¾ inch starter strips.

✔ I've included a few more fasteners than you'll use because, if you're like me, you'll lose or bend a few along the way. It's better to have a few extras on hand and save yourself an extra trip to the hardware store.

Cut List

The following sections break down the Warré hive into its individual components and provide instructions on how to cut and build those components.

Lumber in a store is identified by its *nominal* size, which is its rough dimension before it's trimmed and sanded to its finished size at the lumber mill. The actual finished dimensions are always slightly different from the nominal dimensions. For example, what a lumberyard calls *1 inch x 6 inch lumber* is in fact ¾ inch x 5½ inch, and 2 inch x 3 inch lumber is actually 1½ inch x 2½ inch. In the following sections, each Material column lists nominal dimensions, and each Dimensions column lists the actual, final measurements.

This design involves making rabbet cuts. See Chapter 4 for detailed instructions on making all these cuts.

Hive bottom and stand

Quantity	Material	Dimensions	Notes
4	2" x 3" knotty pine	3⅞" x 2½" x 1½"	These are the leg posts that elevate the hive off the ground.
1	⅜" exterior plywood	16⅛" x 6⁵⁄₁₆" x ⅜"	This is the bees' landing board.
1	⅜" exterior plywood	13¼" x 13¼" x ⅜"	This is the floor. Cut a 4¹¹⁄₁₆" x 4¹³⁄₁₆" notch centered along one edge of the floor board. This is the entrance to the hive.

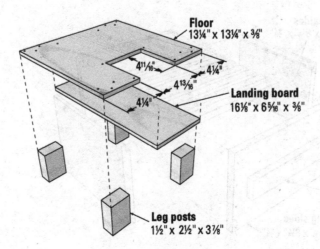

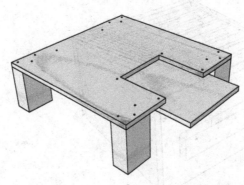

Illustration by Felix Freudzon, Freudzon Design

Hive boxes

Quantity	Material	Dimensions	Notes
8	1" x 10" knotty pine	13⁵⁄₁₆" x 8¼" x ¾"	These are the long sides.
8	1" x 10" knotty pine	11¹³⁄₁₆" x 8¼" x ¾"	These are the short sides. Rabbet a cut ⅜" wide by ⅜" deep along one entire inside top edge. This is the ledge on which the top bar frames sit (frame rest).
8	Use the leftover lumber from your 1" x 10" knotty pine	10" x 2" x ¾"	These are the hand rails. Each hive box has two (you make four hive boxes total, hence eight rails).

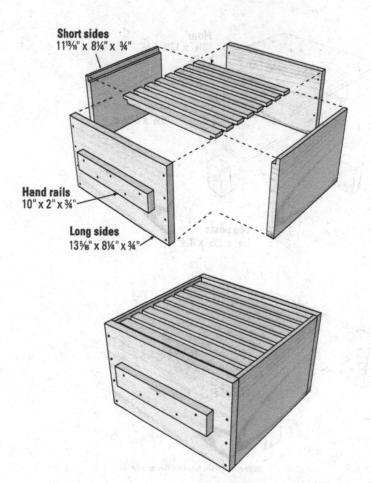

Short sides
11¹³⁄₁₆" x 8¼" x ¾"

Hand rails
10" x 2" x ¾"

Long sides
13⁵⁄₁₆" x 8¼" x ¾"

Illustration by Felix Freudzon, Freudzon Design

Short side (detail)

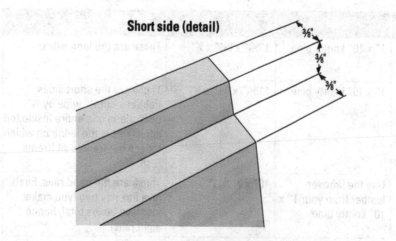

³⁄₈"

³⁄₈"

³⁄₈"

Illustration by Felix Freudzon, Freudzon Design

Top bars

Quantity	Material	Dimensions	Notes
32	1" x 10" knotty pine	12⅜" x 1" x ¾"	Rabbet a cut ⁷⁄₁₆" deep by ⅝" long at each end of the top bars. Cut a saw kerf on the bottom side centered along the entire length, ⅛" wide by ¼" deep (you'll place a starter strip of wood into this groove).
32	³⁄₃₂" balsa wood	10" x ¾" x ³⁄₃₂"	These are the starter strips that you glue into the kerf cut groove.

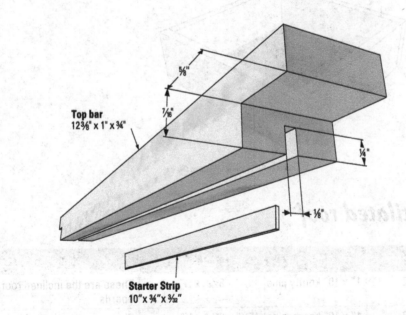

5/8"

7/16"

Top bar
12⅜" x 1" x ¾"

¼"

⅛"

Starter Strip
10"x ¾"x ³⁄₃₂"

Illustration by Felix Freudzon, Freudzon Design

Quilt box

Quantity	Material	Dimensions	Notes
2	1" x 10" knotty pine	13⁵⁄₁₆" x 3¹⁵⁄₁₆" x ¾"	These are the long sides.
2	1" x 10" knotty pine	11¹³⁄₁₆" x 3¹⁵⁄₁₆" x ¾"	These are the short sides.

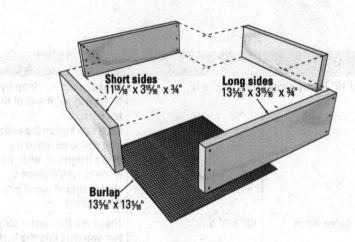

Short sides
$11^{13}/_{16}$" x $3^{15}/_{16}$" x ¾"

Long sides
$13^{5}/_{16}$" x $3^{15}/_{16}$" x ¾"

Burlap
$13^{5}/_{16}$" x $13^{5}/_{16}$"

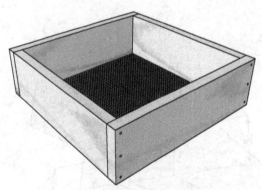

Illustration by Felix Freudzon, Freudzon Design

Ventilated roof

Quantity	Material	Dimensions	Notes
2	1" x 10" knotty pine	19¾" x $8^{5}/_{16}$" x ¾"	These are the inclined roof boards.
2	1" x 10" knotty pine	$15^{3}/_{8}$" x $8^{5}/_{16}$" x ¾"	These are the roof gables. Along the designated top edge of each gable, make a mark with a pencil 6⅞" in from the outside edges. Measuring from the top edge, make a pencil mark 2" down on each side edge. Make an angled cut from marks on the top edge to the marks on the side edge. Do this on both sides of each roof gable. This creates the correct pitch for the gables. It's helpful to refer to the following figure.

Quantity	Material	Dimensions	Notes
2	1" x 10" knotty pine	13⅞" x 4¾" x ¾"	These are the short sides.
1	1" x 10" knotty pine	19¾" x 3" x ¾"	This is the roof ridge board.
1	⅜" exterior plywood	15⅜" x 13⅛" x ⅜"	This is the inner cover board.

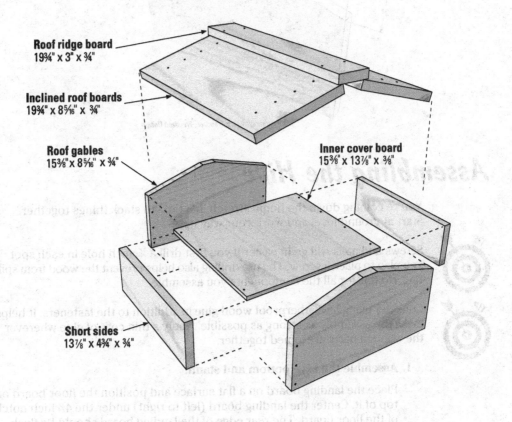

Roof ridge board
19¾" x 3" x ¾"

Inclined roof boards
19¾" x 8⁵⁄₁₆" x ¾"

Roof gables
15⅜" x 8⁵⁄₁₆" x ¾"

Inner cover board
15⅜" x 13⅛" x ⅜"

Short sides
13⅞" x 4¾" x ¾"

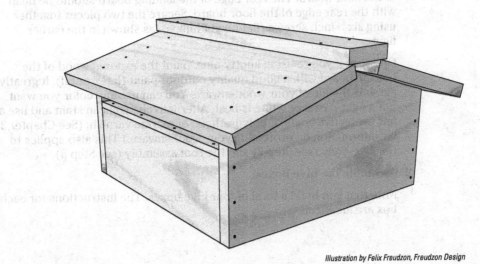

Illustration by Felix Freudzon, Freudzon Design

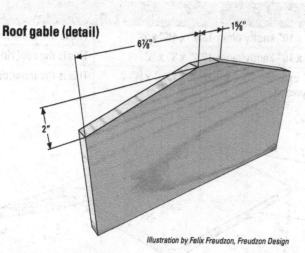

Roof gable (detail)

6⅞"

1⅝"

2"

Illustration by Felix Freudzon, Freudzon Design

Assembling the Hive

You're coming down the home stretch. It's time to stack things together. Start at ground level and work your way up.

Screws and nails will go in easier if you first drill a ⁷⁄₆₄ inch hole in each spot you plan to place a screw. The pre-drilling also helps prevent the wood from splitting. Do this for all the components you assemble.

Consider using a weatherproof wood glue in addition to the fasteners. It helps make the assembly as strong as possible. Apply a thin coat of glue wherever the wooden parts are joined together.

1. **Assemble the hive bottom and stand.**

   Place the landing board on a flat surface and position the floor board on top of it. Center the landing board (left to right) under the 4¾ inch notch of the floor board. The rear edge of the landing board should be flush with the rear edge of the floor board. Secure the two pieces together using six ⅝ inch screws (flat-head Phillips), as shown in the earlier figure.

   Optional: If you're using knotty pine, paint the *exposed* wood of the bottom board with a good quality outdoor paint (latex or oil). It greatly extends the life of your woodenware. You can use any color you want, but a light pastel or white is best. Alternatively, you can stain and use a few protective coats of polyurethane or marine varnish. (See Chapter 3 for information on protecting your woodenware.) This also applies to the hive body (see Step 2) and the roof assembly (see Step 5).

2. **Assemble the hive boxes.**

   Note that you build a total of four hive boxes. The instructions for each box are identical.

Affix the two long sides to the two short sides by hammering one 6d galvanized nail into each of the hive's four edges. Hammer the nails only halfway in to make sure everything is square and fits properly — you have no room for adjustment after you drive these nails (and the remaining 12) all the way in!

Use a carpenter's square to make sure the box stays square as you assemble the hive body. (See Chapter 4 for information on keeping things square.)

When everything looks okay, you can hammer the four nails all the way in, and do the same with the remaining 12 nails.

Now use the #6 x 1⅜ inch galvanized deck screws to attach the two hand rails to opposite sides of the hive box. Choosing which sides get the handles is up to you. Center the hand rails left to right and top to bottom. Use five screws per hand rail, spaced and staggered by eye to prevent splitting the wood, as shown in the earlier figure.

In place of using wooden hand rails, you can attach flush-mounted, galvanized (or stainless steel) handles to the hive box. You can find these handles in hardware stores or marine supply stores. They provide you with an excellent grip.

Check all sides to make certain that all the nails and screws are in place.

3. Assemble the top bars.

You'll assemble a total of 32 top bars. Assembly simply consists of gluing a thin strip of balsa wood into the kerf cut groove. This is the starter strip that gives the bees a starting point to build honeycomb on the top bars. For each bar, center the starter strip into the kerf cut and glue in place using weatherproof wood glue. Let the glue dry before proceeding.

Melt ½ pound of beeswax over low electric heat or in a double boiler. Use a disposable brush to coat the starter strips with a thin coat of beeswax. This further encourages the bees to get started making comb.

Never melt beeswax using an open flame! Beeswax is highly flammable.

Now place the top bars into the hive body. The bars rest on the top edge of the hive body and are butted side by side, like the wooden bars of a marimba.

Never paint your top bars because that could be toxic to your bees. Leave all interior parts of any hive unpainted, unvarnished, and all-natural.

4. Assemble the quilt box.

A quilt box provides insulation to the hive. Affix the two long sides to the two short sides by hammering one 6d galvanized nail into each of the hive's four edges. Hammer them in only halfway to make sure everything is square and fits properly.

Use a carpenter's square to make sure the box stays square as you assemble it.

When everything looks okay, you can hammer the four nails all the way in and do the same with eight additional nails. I use three nails (evenly spaced by eye) in each of the four corners (as shown in the earlier figure).

Using a staple gun, attach the sheet of burlap to what you designate as the bottom of the quilt box. This breathable barrier holds the insulation material in place while allowing for air circulation and ventilation. Use as many staples as you deem necessary. I space my staples about 2 inches apart.

Finally, loosely fill the quilt box with insulation material (such as dry leaves, straw, or natural wood chips).

5. Assemble the ventilated roof.

Affix the peaked gables to the two sides by hammering one 6d galvanized nail halfway into each of the gables' four edges.

Use a carpenter's square to make sure the assembly stays square, and then hammer the four nails all the way in and do the same with eight additional nails. I use three nails in each gable's four corners (evenly spaced by eye).

Now set the inner cover board on top of the edges of the short sides, and attach using four evenly spaced 6d nails on each side. Precise spacing isn't critical — do it by eye.

You nail the two inclined roof boards to the gables. Use four evenly spaced nails per edge. Align the roof boards such that there's a 1⅛ inch ventilation gap at the peak of the roof.

Finally, attach the ridge rail to the peak (covering the ventilation gap). Use two 6d nails at each end of the ridge rail.

6. Stack all the pieces together to create the Warré hive (see the following figure).

Place the hive bottom and stand on the ground. This serves as the landing board and bottom of the hive. It also elevates the hive off the damp ground and improves air circulation.

Stack the four hive bodies (filled with the top bars) on top of the hive bottom and stand. These boxes are where the bees build their comb, raise their brood, and store their pollen and honey.

Now add the quilt box. It provides a layer of insulation for the colony. Make sure you've loosely filled the box with dry leaves, straw, or natural wood chips.

Finish everything off with the ventilated roof, which, in addition to ventilation, protects the colony from the elements.

Hooray! You're ready for your bees.

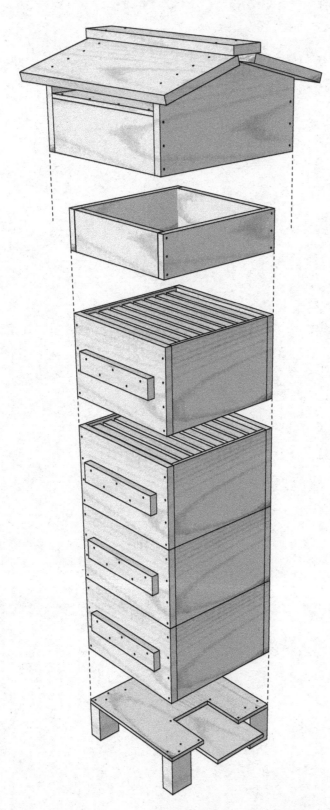

Illustration by Felix Freudzon, Freudzon Design

Chapter 9

The British National Hive

Illustration by Felix Freudzon, Freudzon Design

Although the Langstroth hive in Chapter 10 is the most popular hive in the United States, the number one hive in the United Kingdom is the British National hive (BNH). That's been the case for many years.

An advantage of this hive is that the bees build honeycomb into frames that you can remove, inspect, and move about with ease. The brood chamber holds 11 deep frames. Each deep frame has a comb area of roughly 100 square inches per side. In addition, each shallow honey super holds 11 shallow frames with a comb area of roughly 68 square inches per side.

Like the Langstroth, because of its size and expandability, this hive does well in climates similar to that of the UK (namely, where summers are hot and winters can be cold, the average annual temperature is between 45 and 65 degrees, and minimum temperatures fall between 15 and 35 degrees). I love the fact that the roof has four ventilation holes because I strongly believe that good ventilation plays a critical role in keeping colonies healthier and more productive.

Because the BNH is mostly unique to the UK, you may have a devil of a time finding accessories, frames, and foundation outside of the UK. You can always order from UK-based suppliers, but if that means overseas shipping for you, you may want to keep that in mind as you decide whether this is the right hive for you.

Vital Stats

- ✔ **Overall size:** 20⅜ inches x 20⅜ inches x 24⅝ inches.

- ✔ **Capacity:** This design consists of modular, interchangeable hive parts, so you can add more shallow honey supers as the colony grows and honey production increases. The capacity for bees and honey is unlimited.

- ✔ **Type of frame:** Because frames and foundation for this hive are typically only available from UK-based suppliers, I've designed a hybrid frame that you can make yourself. It's part top bar and part conventional BNH frame. These frames have a wooden top and the support of a bottom rail and sides, but they don't use wired beeswax foundation (which is traditionally used in the BNH design but isn't readily available in the United States). Instead, a thin starter strip of balsa wood allows the bees to build their own comb freestyle and without restrictions on cell size. That's a great all-natural option for the bees, but the frames of capped honey are more delicate than frames made with conventional wired foundation. Therefore, you have to be very careful when using an extractor. On the other hand, you can readily harvest cut-comb honey or crush the combs to extract the honey.

For extra strength, you can zig-zag some support wiring through the holes in the frames of the side bars to provide extra support as the bees build their comb.

If you decide to order ready-made BNH frames and foundation, I suggest that you contact E. H. Thorne (Beehives) Ltd. Find this English company on the web at www.thorne.co.uk. The frames and foundation it sells should work fine with this design.

✔ **Universality:** Given this design's wide popularity in the UK, the beekeeper using the BNH has all kinds of options for purchasing extras (such as replacement parts and accessories). But you'll likely have to use a supplier in the UK to get these parts. I'm not aware of any beekeeping suppliers in the United States that carry BNH parts and accessories.

For the version of the BNH in this chapter, I've designated the bee space at the *top* of the hive, which is more typical for American beekeepers. Having the correct bee space means the bees won't glue parts together with propolis or burr comb (see Chapter 4 for more information on bee space). *Bottom* bee space is preferred in the UK, however. Because this design uses top bee space, not all commercially available parts and accessories from the UK will be compatible. So if you order commercially available items for this hive, be sure to ask the vendor if what you order is compatible with *top* bee space hives. (*Note:* If you elect to order BNH frames and foundation from the UK, they will work fine with hives using *either* top or bottom bee space.)

✔ **Degree of difficulty:** This is a pretty straightforward design. However, the tin work involved with the aluminum flashing material used on the roof can be tricky. Bending the corners, like folding the corners of sheets on a bed, takes a little patience and practice.

✔ **Cost:** Using scrap wood (if you can find some) would keep material costs of this design minimal, but even if you purchase the recommended wood, hardware, and fasteners, you can likely build this hive (frames and all) for less than $175 (and even less if you use pine lumber in place of the cedar).

Materials List

The following table lists what you'll use to build your BNH and the frames within it. Feel free to change out some materials to suit your needs or take advantage of materials you have on hand. (See Chapter 3 for more on selecting materials.)

	Lumber		Hardware		Fasteners
3	8' lengths of 1" x 6" clear pine lumber		Weatherproof wood glue	200	#6d x 2" galvanized nails
1	8' length of 1" x 8" cedar		½ pound pure beeswax for melting	320	⁵⁄₃₂" x 1⅛" flat-head, diamond-point wire nails
1	10' length of 1" x 8" cedar	1	roll of 24" wide aluminum flashing (usually comes in a 10' length). You can use the extra material to build additional hives.	20	#6 x 1⅝" deck screws, galvanized, #2 Phillips drive, flat-head with coarse thread and sharp point

(continued)

Lumber	Hardware	Fasteners
1 10' length of 1" x 10" cedar	Small piece of ⅛" (#8) hardware cloth (5" x 5" will do the trick)	25 #8 x ½" lath screws, galvanized, #2 Phillips drive, flat-head with sharp point
1 8' length of 2" x 6" stud spruce or fir	Optional: 2 quarts latex or oil exterior paint (white or any light color), exterior polyurethane, or marine varnish	
1 12" x 36" x ³⁄₃₂" balsa wood sheet (available from hobby supply stores)		
1 8' length of ¾" x 6" cedar decking		
1 2' x 4' sheet of ¾" thick exterior plywood		
1 2' x 4' sheet of ¼" thick lauan plywood		

Here are a couple of notes about the materials for your British National hive:

✔ In the UK, British National hives are traditionally made from cedar, so that's what I've specified for these plans. You can always swap out the cedar for the more economical knotty pine option.

✔ Pure beeswax for melting is available at arts and crafts stores and from beekeeping supply vendors; you can find many vendors online. Here are some to consider, in no particular order:

  • www.naturalcrafts.glorybee.com

  • www.brushymountainbeefarm.com

  • www.dadant.com

  • www.hiveharvest.com

✔ I've included a few more fasteners than you'll actually use, just in case you lose or bend a few along the way.

Cut List

The following sections break down the BNH into its individual components and provide instructions on how to cut and build those components.

Lumber in a store is identified by its *nominal* size, which is its rough dimension before it's trimmed and sanded to its finished size at the lumber mill. The actual finished dimensions are always slightly different from the nominal dimensions. For example, what a lumberyard calls *1 inch x 8 inch lumber* is in fact ¾ inch x 7½ inch, and *1 inch x 10 inch lumber* is in fact ¾ inch x 9½ inch. In the following tables, each Material column lists nominal dimensions, and each Dimensions column lists the actual, final measurements.

The frames and other pieces of this hive require a number of tricky cuts, including dado and rabbet cuts. See Chapter 4 for detailed instructions on making these cuts.

Floor

Quantity	Material	Dimensions	Notes
2	1" x 10" cedar	18⅛" x 2" x ¾"	These are the side rails. Dado ¾" wide by ⅜" deep along the entire length of each side rail. Rabbet one end of each side rail ¾" wide by ⅜" deep. Be sure to rabbet opposite ends.
1	1" x 8" cedar	17⅜" x 2" x ¾"	This is the rear rail. Dado ¾" wide by ⅜" deep along the entire length.
1	1" x 8" cedar	16⅝" x ⅞" x ¾"	This is the entrance reducer. Cut one notch centered on one side of the entrance reducer (⅜" deep by 4" wide).
1	¾" exterior plywood	17⅜" x 17¾" x ¾"	This is the actual "floor."

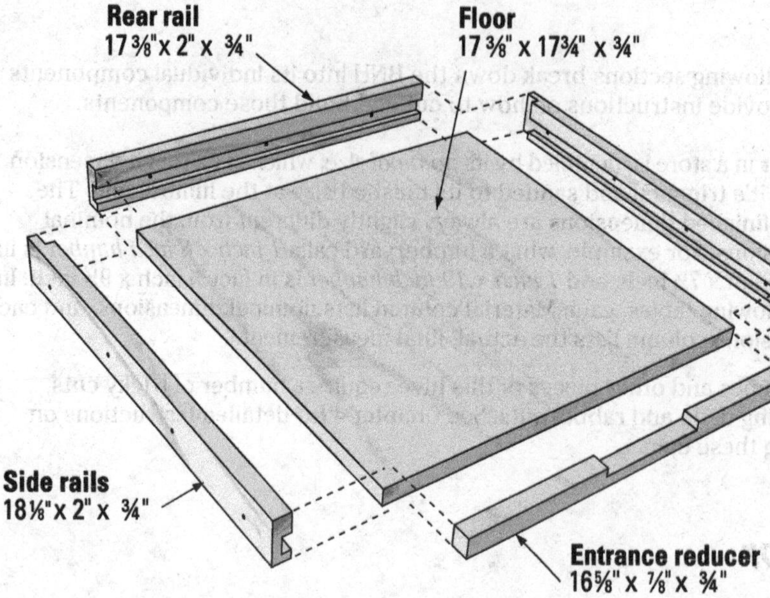
Rear rail
17 3⁄8" x 2" x 3⁄4"
Floor
17 3⁄8" x 17 3⁄4" x 3⁄4"
Side rails
18 1⁄8" x 2" x 3⁄4"
Entrance reducer
16 5⁄8" x 7⁄8" x 3⁄4"

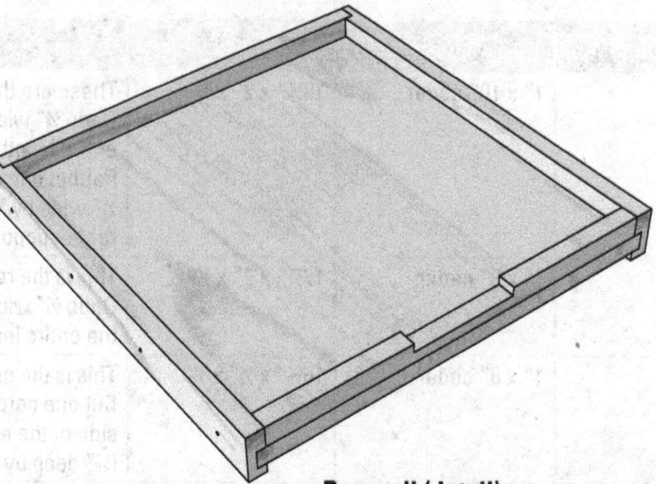

Side rail (detail)

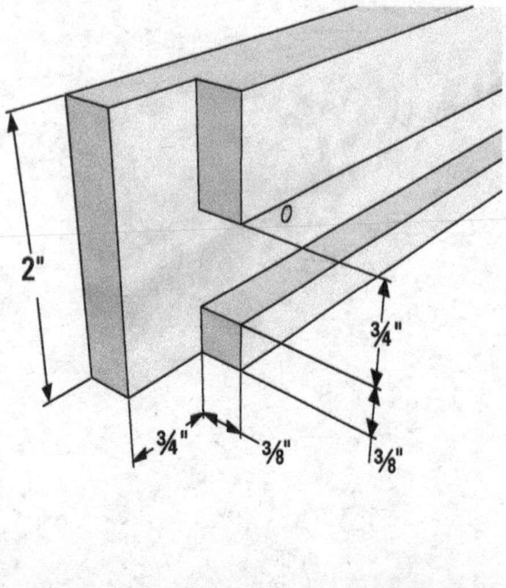
2"
3⁄4"
3⁄4"
3⁄8"
3⁄8"

Rear rail (detail)

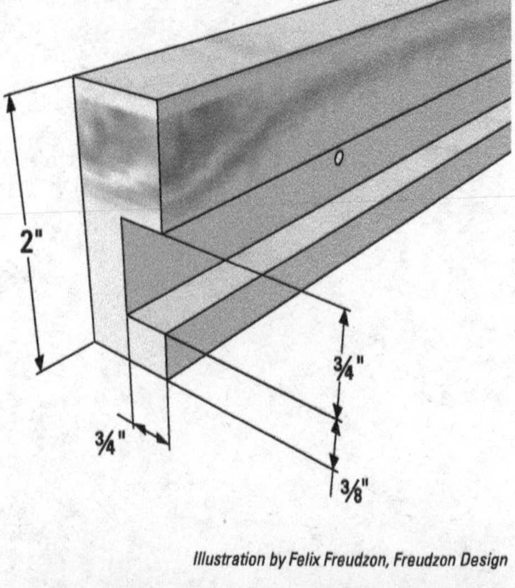
2"
3⁄4"
3⁄4"
3⁄8"

Illustration by Felix Freudzon, Freudzon Design

Brood chamber

Quantity	Material	Dimensions	Notes
4	⅝" x 6" cedar decking	18⅛" x 1½" x 1"	Two of these are the upper hand rails and the other two are the lower drip rails. Note the special cuts I describe in the later section "Making tricky cuts for hand and drip rails."
2	1" x 10" cedar	18⅛" x 8⅛" x ¾"	These are the long sides. Dado a groove at each end, 1" in from the outer edge, ¾" wide, and ⅜" deep. Cut two notches at both ends, ½" from bottom (and top edge), ½" wide, and ½" deep.
2	1" x 10" cedar	17⅞" x 8³⁄₁₆" x ¾"	These are the short sides. Rabbet ¼" wide by ½" deep along one entire outside top edge. This becomes the ledge on which the top bar frames sit (frame rest).

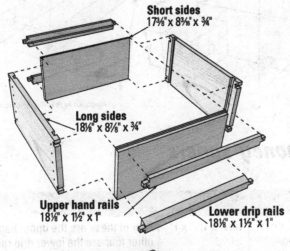

Short sides
17⅞" x 8³⁄₁₆" x ¾"

Long sides
18⅛" x 8⅛" x ¾"

Upper hand rails
18⅛" x 1½" x 1"

Lower drip rails
18⅛" x 1½" x 1"

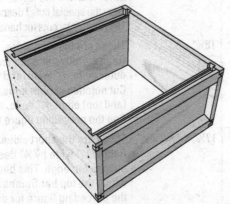

Illustration by Felix Freudzon, Freudzon Design

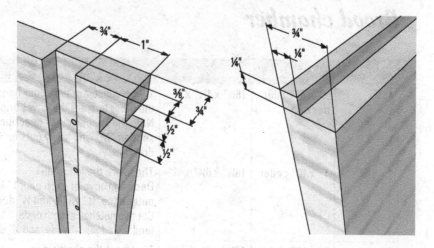

Lower drip rail (detail)　　**Upper hand rail (detail)**

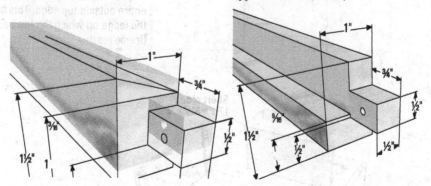

Illustration by Felix Freudzon, Freudzon Design

Shallow honey supers

Quantity	Material	Dimensions	Notes
8	⅝" x 6" cedar decking	18⅛" x 1½" x 1"	Four of these are the upper hand rails and the other four are the lower drip rails. Note the special cuts I describe in the later section "Making tricky cuts for hand and drip rails."
4	1" x 8" cedar	18⅛" x 5⅝" x ¾"	These are the long sides. Dado a groove at each end, 1" in from the outer edge, ¾" wide, and ⅜" deep. Cut notches at both ends, ½" from bottom (and top) edge, ½" wide, and ½" deep. See the preceding figure for details.
4	1" x 8" cedar	17⅜" x 5⁵⁄₁₆" x ¾"	These are the short sides. Rabbet ¼" wide by ½" deep along one entire outside top edge. This becomes the ledge on which the top bar frames sit (frame rest). See the preceding figure for details.

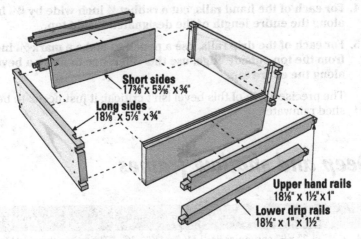

Short sides
17⅜" x 5³⁄₁₆" x ¾"

Long sides
18⅛" x 5⅞" x ¾"

Upper hand rails
18⅛" x 1½" x 1"

Lower drip rails
18⅛" x 1" x 1½"

18⅛" x 18⅛" x 5⅞"

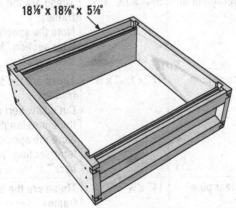

Illustration by Felix Freudzon, Freudzon Design

Making tricky cuts for hand and drip rails

Some fussy cuts are involved with hand and drip rails for the brood chamber and shallow honey supers. Read through the following list and refer to the figures in the earlier "Brood chamber" and "Shallow honey supers" sections to better understand how to make these cuts. You can make all cuts on your table saw.

1. **On each end of each rail, use a ¾ inch dado blade to cut a ½ inch deep rabbet along the designated bottom of each rail.**

   Do this for the hand rails and the drip rails.

2. **Use the ¾ inch dado blade to cut a ½ inch deep rabbet along the designated inside edge of each end of all the rails.**

   Do this for the hand rails and the drip rails.

3. **Use the ¾ inch dado blade to cut a ½ inch deep rabbet along the designated top edge of each end of all the rails.**

   Do this for the hand rails and the drip rails. As a result of these cuts, you'll be left with a ¾ inch x ½ inch x ½ inch tab centered on the designated outside edge of each rail.

4. **For each of the hand rails, cut a rabbet ½ inch wide by $^{15}/_{16}$ inch deep along the entire length of the designated inside top.**

5. **For each of the drip rails, use a pencil to make a mark $^5/_{16}$ inch down from the top outside edge; use this reference to make a beveled cut along the entire top.**

The precise angle of this bevel isn't critical; it just needs to be enough to shed rainwater.

Deep and shallow frames

Quantity	Material	Dimensions	Notes
44	2" x 6" spruce or fir	$5^9/_{16}$" x $1^3/_8$" x $^3/_8$"	These are the side bars for the shallow frames. Note the special cuts I describe in the later section "Making tricky cuts for side bars."
33	1" x 6" clear pine	17" x $1^1/_{16}$" x $^3/_4$"	These are the top bars for all the frames. Cut a saw kerf groove centered along the entire length, $^1/_8$" wide by $^1/_4$" deep. Note the special cuts I describe in the later section "Making tricky cuts for top bars."
33	1" x 6" clear pine	14" x $^3/_8$" x $^3/_4$"	These are the bottom bars for all the frames.
33	$^3/_{32}$" balsa wood	12" x 1" x $^3/_{32}$"	Use a utility knife to cut balsa wood sheets into starter strips.
22	2" x 6" spruce or fir	$8^9/_{16}$" x $1^3/_8$" x $^3/_8$"	These are the side bars for the deep frames. Note the special cuts I describe in the later section "Making tricky cuts for side bars."

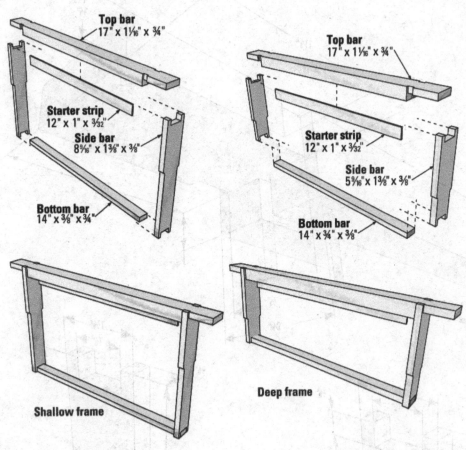

Top bar
17" x 1¹⁄₁₆" x ¾"

Starter strip
12" x 1" x ³⁄₃₂"

Side bar
8⁵⁄₁₆" x 1³⁄₈" x ³⁄₈"

Bottom bar
14" x ³⁄₈" x ¾"

Top bar
17" x 1¹⁄₁₆" x ¾"

Starter strip
12" x 1" x ³⁄₃₂"

Side bar
5⁹⁄₁₆" x 1³⁄₈" x ³⁄₈"

Bottom bar
14" x ¾" x ³⁄₈"

Shallow frame

Deep frame

Illustration by Felix Freudzon, Freudzon Design

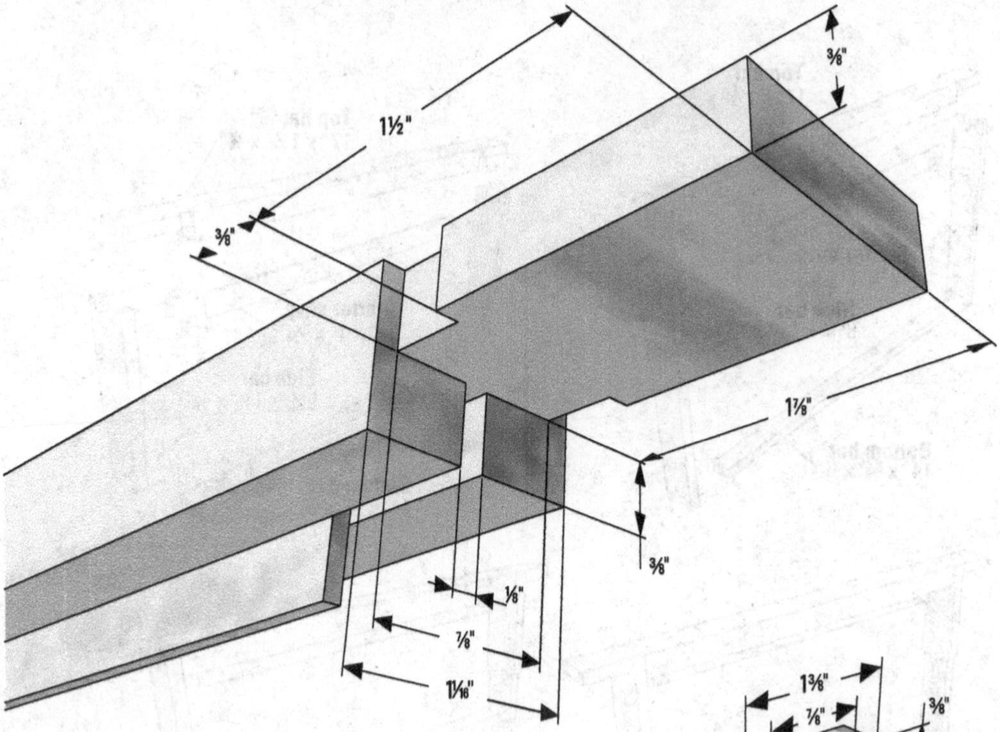
1½"
⅜"
⅜"
1⅞"
⅜"
⅛"
⅞"
1 1/16"

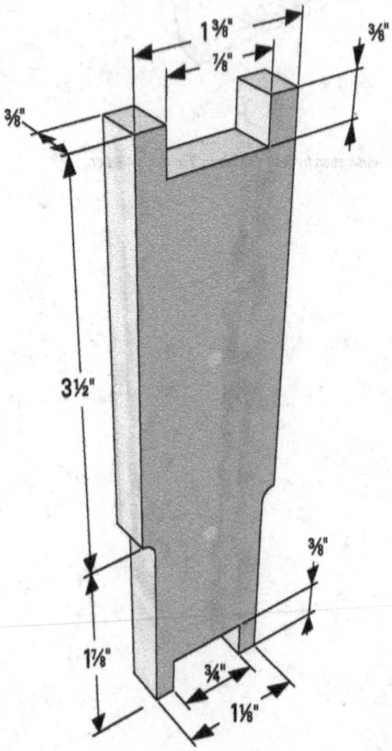
1⅜"
⅞"
⅜"
⅜"
3½"
⅜"
1⅞"
¾"
1⅛"

Shallow side bar

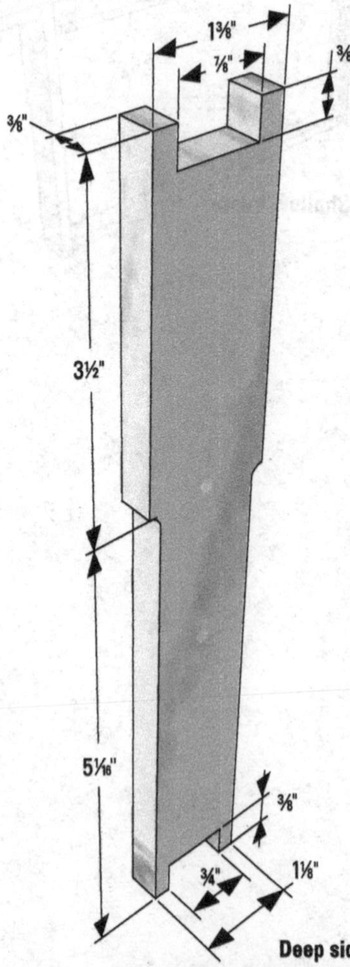
1⅜"
⅞"
⅜"
⅜"
3½"
5 1/16"
⅜"
¾"
1⅛"

Deep side bar

Illustration by Felix Freudzon, Freudzon Design

Making tricky cuts for side bars

Side bars have a wide profile at the top and taper to a narrower profile at the bottom. This tapered shape provides the correct distance between frames and allows for the proper bee space around and in between the frames so that bees can travel freely (and so they don't glue the frames together; see Chapter 4 for more about bee space). The top of each end bar has a notch to accommodate the top bar, and the bottom has a notch to accommodate the bottom bar. Follow these steps and refer to the figure in the preceding section, "Deep and shallow frames," to make the cuts for side bars. The basic steps are identical for deep and shallow side bars. Make these cuts using your table saw or with a table router if you have one.

I find it easiest to make frames assembly-line fashion — I set up my work space for making one particular cut and then repetitively make that cut on *all* the pieces that call for it. For example, I cut out all the top notches on all the side bars before readjusting my tools and measurements and moving on to the bottom notches.

1. **Create the taper by removing ⅛ inch of material from each vertical edge of the bar.**

   Note that the lower portion of the side bar is narrower than the upper portion. For both the deep and shallow side bars, the taper cut starts 3½ inches down from the top.

2. **Cut a notch ⅛ inch wide by ⅜ inch deep at the top of the side bar.**

   The top bar snaps into this notch when you assemble the frame.

3. **Cut a notch ¾ inch wide by ⅜ inch deep at the bottom of the side bar.**

   The bottom bar snaps into this notch when you assemble the frame.

Making tricky cuts for top bars

Follow these steps to make tricky cuts for top bars (and, because a picture is worth a thousand words, be sure to refer to the figure in the earlier section, "Deep and shallow frames"):

1. **Cut a kerf ⅛ inch wide by ⅜ inch deep along the entire long length of the designated underside of the top bar.**

2. **Cut a vertical notch ⅜ inch wide by ³⁄₃₂ inch deep on both sides and at each end of the top bar.**

   The notch starts 1½ inches back from the ends of the bar. When you assemble the frames, you insert the top of the side bars into these notches.

3. **Working from the underside of each end of the bar, make a ⅜ inch deep by 1⅛ inch wide rabbet.**

   This creates the tabs at each end of the top bar.

Crown board

Quantity	Material	Dimensions	Notes
4	1" x 6" clear pine	18" x ¾" x ¾"	These are the long side rails. Dado ¼" wide by ⅜" deep along entire length, ⅛" from edge.
2	1" x 6" clear pine	17¼" x ¾" x ¾"	These are the short side rails. Dado ¼" wide by ⅜" deep along entire length, ⅛" from edge. Rabbet opposite ends, ¾" wide by ⅜" deep.
1	¼" lauan plywood	17¼" x 17¼" x ¼"	This is the top. Cut two 4½" x 1⅝" ovals in the top. Center the ovals left to right. Position one 6" from the front edge of the plywood board and the other 2⅝" from the front edge of the plywood board. These holes serve for ventilation and can also be used to accommodate a Porter bee escape at honey harvest time.

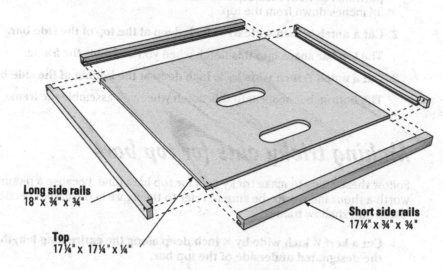

Long side rails
18" x ¾" x ¾"

Short side rails
17¼" x ¾" x ¾"

Top
17¼" x 17¼" x ¼"

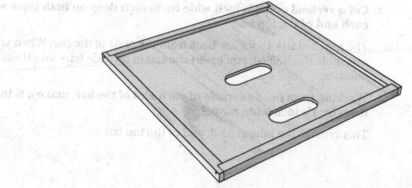

Illustration by Felix Freudzon, Freudzon Design

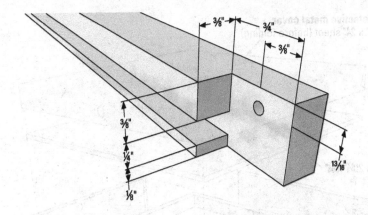

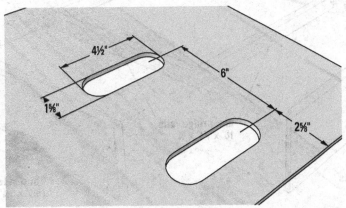

Illustration by Felix Freudzon, Freudzon Design

Roof

Quantity	Material	Dimensions	Notes
4	1" x 6" clear pine	18" x 1¼" x ¾"	Two of these are the long sides of the inner ridge rail, and the other two are the short sides of the inner ridge rail.
2	1" x 10" cedar	20¼" x 5¾" x ¾"	These are the long side panels. Rabbet the edge of each corner ⅜" deep by ¾" wide (see the following figure).
2	1" x 10" cedar	19½ x 5¾" x ¾"	These are the short side panels.
1	¾" exterior plywood	20¼" x 20¼" x ¾"	This is the top.
1	24" aluminum flashing	24" x 24"	This is the protective metal cover. Use tin snips to cut the flashing to size. Wrap flashing over the top of the assembled roof. There will be a flap that folds over the top edge. Fold the corners to avoid sharp edges.

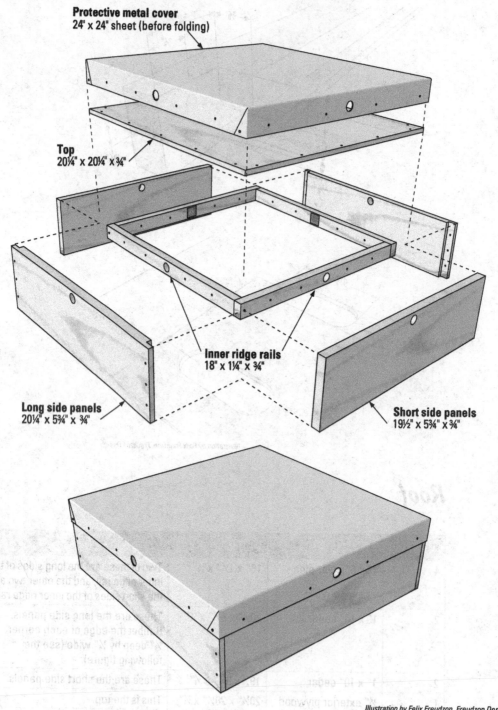

Protective metal cover
24" x 24" sheet (before folding)

Top
20¼" x 20¼" x ¾"

Inner ridge rails
18" x 1¼" x ¾"

Long side panels
20¼" x 5¾" x ¾"

Short side panels
19½" x 5¾" x ¾"

Illustration by Felix Freudzon, Freudzon Design

Assembling the Hive

Putting together the components of your BNH is pretty much just stacking one component on top of another (like a skyscraper). But there is, of course, a correct sequence. Understanding each element's purpose is helpful in understanding the sequence in which you build and stack them. You start at the bottom and work your way up.

Throughout this project, use a weatherproof wood glue in addition to the fasteners. It helps make the hive parts as strong as possible. Apply a thin coat of glue wherever the wooden parts are joined together.

Nails will go in easier if you first drill a ⁷⁄₆₄ inch hole in each spot you plan to place a nail. This drilling also helps prevent the wood from splitting. Use this little trick for all the hive parts you assemble.

1. **Assemble the floor.**

   Place the short rail on your worktable with the dado side up. Position the plywood floor into the dado groove. You can use either end of the plywood floor as the *rear* of the bottom board.

   Some edges of plywood tend to look a little nicer than others. I like to hide the ugly edge in the dado, leaving the nicer looking edge as the exposed edge.

   Place the long rails on both sides of the plywood floor, inserting the plywood into the dados.

   Be certain that the dado in all rails faces the same way (the dado isn't centered along the rail). Otherwise, you'll have a seriously lopsided bottom board!

   Check the alignment and fit of all the rails with the floor, and then place one of the #6d x 2 inch galvanized nails halfway into the center of each of the three rails (the nails go through the rails and into the edges of the plywood). Don't hammer them all the way in quite yet. First make sure everything fits properly; you have no room for adjustment when all the nails are in! When the fit looks good, hammer these nails all the way in, plus three additional nails per side rail and two additional for the rear rail.

   The entrance reducer remains loose and is placed in the entrance of the hive to control ventilation and prevent *robbing,* that dreaded occurrence when bees from neighboring hives invade another hive to steal all the honey. The entrance reducer makes the opening of a weak hive smaller and thus easier to defend from robbing neighbors. The entrance reducer is typically not used year-round. For info on using an entrance reducer, see Chapter 2.

2. **Assemble the brood chamber.**

Insert the edges of the short sides into the dado grooves cut into the long sides. Affix them together by hammering one #6d galvanized nail into each of the brood box's four edges. The rabbeted edges of the two short sides represent the top of the brood chamber (the frames will rest on the ledge the rabbet creates). Hammer the nails only *halfway* in to make sure everything is square and fits properly.

Use a carpenter's square to make sure the box stays square as you assemble the brood chamber. (See Chapter 4 for info on keeping things square.)

When everything looks okay, you can hammer the nails all the way in and add an additional 4 evenly spaced nails per edge. Check all sides to make certain that all 20 nails are in place.

3. **Assemble the shallow honey supers.**

You assemble *two* shallow honey supers. The instructions for each are identical. Assembly is similar to what you did for the brood chamber in the preceding step.

Insert the edges of the short sides into the dado grooves cut into the long sides. Affix them together by hammering one #6d galvanized nail into each of the super's four edges. The rabbeted edges of the two short sides represent the top of the super (the frames will rest on the ledge the rabbet creates). Hammer the nails only *halfway* in to make sure everything is square and fits properly.

Use a carpenter's square to make sure the box stays square as you assemble the shallow honey super.

When everything looks okay, you can hammer the nails all the way in and add an additional 2 evenly spaced nails per edge. Check all sides to make certain that all 12 nails are in place.

You're done making one shallow honey super. Follow the same procedure to build the second super.

4. **Assemble the deep and shallow frames.**

You'll assemble 11 deep frames and 22 shallow frames. The instructions are identical for both deep frames and shallow frames. Only the vertical height differs.

Place a top bar on your work surface with the ⅛ inch kerf cut facing up.

Insert the wider end of the two side bars (8⁹⁄₁₆ inch length if you're building deep frames or 5⁵⁄₁₆ inch length if you're assembling shallow frames) into the slots at either end of the top bar.

Insert the bottom bar into the slots at the narrow ends of the side bars. Make sure the frame assembly is square.

Now nail all four pieces together. Use a total of six 1⅛ inch flat-head nails per frame (two for each end of the top bar and one at each end of the bottom bar).

Repeat these steps until you've assembled all your frames. Jolly good!

Don't be tempted to use any shortcuts when you build frames. Frames undergo all kinds of abuse and stress, so their structural integrity is vital. Don't skimp on the nails or settle for a bent nail that's partially driven home. There's no cheating when it comes to assembling frames!

Never paint your frames; that could be toxic to your bees. Always leave all interior parts of any hive unpainted, unvarnished, and all-natural.

5. **Install and prime the frames' starter strips.**

Take a balsa wood starter strip and glue it (centered) into the ⅛ inch kerf cut groove that's on the bottom of the top bar. Repeat this step for the remaining 32 frames. Let the glue dry completely before proceeding to the next step.

Melt the ½ pound of beeswax over low electric heat or in a double boiler. Use a disposable brush to coat all the starter strips with a thin coat of beeswax. This priming encourages the bees to get started making comb.

Never melt beeswax using an open flame! Beeswax is highly flammable.

Now place 11 frames into the brood box and 11 frames into each of the two shallow honey supers (33 frames in all). The frames rest on the rabbeted ledge. Distribute the frames evenly within the brood box and supers.

6. **Assemble the crown board.**

Position the ¼ inch plywood cover insert into the dado of the long rails and the short rails. It's kind of like putting a picture frame together.

Be certain that all rails have the thick or *thin* lip of the groove facing the same way. Otherwise, you'll have a seriously lopsided crown board!

Check the alignment and fit and insert a #6 x 1⅝ inch galvanized deck screw halfway into each of the four corners on the long rails. When everything is square and fits properly, screw them in all the way. Keep in mind that if you cut the plywood insert square it squares up the frame nicely.

Don't paint the crown board. Leave it natural and unfinished, as with all the internal parts of any beehive.

7. **Assemble the panels of the roof.**

Loosely assemble all four side panels of the roof together as a "test fit" (you're building a box). The short sides fit into the rabbeted edges of the long sides.

When assembling the roof, having a *stop* on your worktable that you can push against while inserting screws is helpful. A short piece of 2x4 lumber clamped to the table serves as a good stop.

Place one of the long panels of the loosely assembled roof against the stop. Using the galvanized deck screws, insert three screws into each corner of the long panel at the opposite end of the assembly. Reverse the entire cover end to end and screw the corners of the other long panel in a similar manner. Make certain that the entire assembly remains snug, tight, and square as you do this.

Now use the deck screws to secure the plywood to the top of the assembly. Drive the screws through the plywood top and into the top edges of the four side panels. Use five evenly spaced screws along each side.

Optional: Paint the *exposed* exterior wood of the roof with a good quality outdoor paint (latex or oil). Doing so greatly extends the life of your woodenware. You can use any color you want, but white or a light pastel is best. Do *not* paint the inside of the top or the aluminum flashing you add in the next step. Alternatively, you can stain and polyurethane the exterior wood. (See Chapter 3 for information on protecting your woodenware.)

8. **Add the aluminum flashing and the inner ridge frame to the roof.**

Center the 24-inch-x-24-inch aluminum flashing evenly on the top and bend the flashing over the edges of the box. This creates a lip all around the top edge. Do this to all four sides. Bend and fold the corners like you're making the corners of a bed. The flashing is thin and fairly easy to work with. Use a rubber mallet to coax the corners flush and flat.

The edges of aluminum flashing are very sharp. Use caution when handling flashing to avoid cutting yourself and consider using work gloves.

Affix the folded edges of the flashing to the outer cover using the #8 x ½ inch lath screws. I use four evenly spaced screws per side, plus an extra screw to secure each folded corner.

If you don't feel up to the task of fitting the aluminum flashing on the roof, you have another option. Use #15 pound felt tar paper roofing material instead. You can staple it in place (rather than using screws). It's a lot easier to work with, and some would argue that it results in a quieter roof when it rains and therefore is less stressful to the bees.

Assemble the inner ridge frame using one deck screw in each corner. These are simple butt joints. Make certain the assembly is square.

Place the roof assembly upside down on a flat work surface with the metal roof resting on the work surface. Insert the inner ridge frame into the roof assembly and secure it to the walls of the roof using four deck screws per rail. Screw through the ridge frame's rails and into the roof's side walls. The ridge frame fits snugly against the underside of the roof's top. The inner ridge frame provides a ventilated air space.

Now use a ½ inch bit and drill ventilation holes all the way through the aluminum flashing, through the roof's walls, and through the inner ridge frame. Drill one hole per side. Center each hole left to right and position each ¼ inch down from the top edge of the roof. Now staple a ¾ inch square of #8 hardware cloth over each of the holes from the *inside* of the roof. That keeps the air flowing and the bees inside.

9. **Stack all the pieces together to create the hive.**

Now it's time to put all the elements together. Place the floor on level ground. The floor keeps the colony off the damp ground and provides for the hive's entrance (where the bees fly in and out).

Consider using an elevated hive stand to raise the hive farther off the ground and make it more accessible to you. (See Chapter 13 for instructions on building an elevated hive stand that works with this hive.)

Next, the brood box goes on top of the floor. The bees raise baby bees and store food for their use in these boxes.

Stack the two shallow honey supers on top of the deep brood box. This is where the bees store honey — the honey you'll be harvesting for yourself. Each shallow super holds 25 to 30 pounds of honey.

When your top honey super is about half-full with capped honey, it's time to build another super and 11 more frames. If you're lucky and the honey flow is heavy, you may ultimately stack three, four, or more supers on your hive. A royal event for your British hive!

Place the crown board on top of the uppermost super. Position it with the flat side down and with the cutout ovals closest to the hive's entrance.

The roof sits on top of it all. It provides ventilation and protection from the elements. The BNH uses a *telescoping cover* design, meaning the cover fits on and over the hive (like a hat). Simply stack the roof on top of your inner cover and that's it — your British National hive is ready for her royal majesty and her subjects!

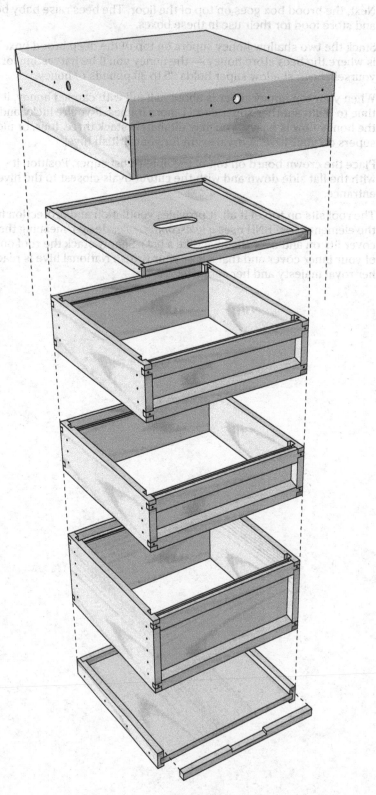

Chapter 10
The Langstroth Hive

Illustration by Felix Freudzon, Freudzon Design

Hands down, the Langstroth hive is the most popular and widely used hive today. Certainly this is the case in the United States and in most developed countries. Invented in 1852 by Reverend Lorenzo Langstroth, the basic design has remained unchanged, which is a testament to its practicality.

The big advantage of this hive is that the bees build honeycomb into frames, which you can remove, inspect, and move about with ease. All the hive's interior parts are spaced exactly ⅜ inches apart, thus providing the correct "bee space" (in other words, the bees won't "glue" parts together with propolis or burr comb; see Chapter 4 for details).

Another advantage: Given its wide popularity, many commercially available parts and accessories sold for beehives are standardized to accommodate this design, meaning the beekeeper using a Langstroth hive has all kinds of options for purchasing extras (such as replacement parts and accessories).

Most bee supply vendors offer ten-frame versions of the Langstroth hive — each hive body holds ten frames across. For decades, that's been the most popular size for the Langstroth. However, as a result of some recent books and publications, the eight-frame version of this hive is gaining popularity. And for good reason — fewer frames means a lighter weight hive body and super, and there's a lot to be said for that! Some of the commercial beekeeping suppliers have branded the eight-frame Langstroth as the _garden hive_.

In this chapter, I include cut sheets for building both the traditional ten-frame version and the increasingly popular eight-frame version. Note that the figures and their respective call-out measurements are only shown for the ten-frame version of this hive. The basic assembly, though, is identical for both versions.

Keep in mind that other than the frames, the parts and accessories for a ten-frame hive are not interchangeable with the parts and accessories for an eight-frame hive. So you need to decide which of these versions of the Langstroth hive is right for you.

I also specify _medium_ honey supers rather than _shallow_ (the super is where the bees store the surplus honey you harvest). I have a strong preference for using medium honey supers over shallow because they hold more honey and yet are still small and light enough to lift off the hive.

The only disadvantage of the Langstroth hive (in terms of building) relates to the use of _finger joints_ (also known as _box joints_ or _comb joints_). These are fairly traditional and what are used on the top quality Langstroth hives available from the commercial beekeeping supply vendors. Although this joint is the strongest and most desirable type of joinery for this hive, making finger joints can be tricky for the neophyte carpenter and requires having the right tools for the job. (For help making finger joints, see Chapter 4.)

Vital Stats

- ✔ **Overall size:** 22 inches x 18 inches x 29¼ inches (ten-frame version); 22 inches x 15¾ inches x 29¼ inches (eight-frame version).

- ✔ **Capacity:** Because this design consists of modular, interchangeable hive parts, you can add extra medium honey supers as the colony grows and honey production increases. The capacity for bees and honey is unlimited, regardless of whether you build the eight- or ten-frame version of the hive.

- ✔ **Type of frame:** This hive uses a Langstroth-style, self-centering frame with beeswax foundation inserts. The ten-frame version has 20 deep frames and 10 medium frames. The eight-frame version has 16 deep frames and 8 medium frames.

✔ **Universality:** Because the Langstroth hive is so widely used around the world, you can easily find replacement parts, gadgets, and add-ons, even for the more recently popular eight-frame version. This stuff is widely available from many beekeeping supply stores (search the web and you'll find dozens of such suppliers). Also, you can easily purchase Langstroth-style frames (and I strongly recommend doing so, as they're fairly tricky to build). When ordering, just specify _deep_ or _medium_ size Langstroth frames and foundation. However, if you want to try building your own frames for this hive, check out Chapter 17.

✔ **Degree of difficulty:** This is a pretty straightforward design. However, two details add a moderate degree of difficulty:

- The fabrication of the finger joints is likely the trickiest part for beginner woodworkers.
- For the tin work involved with the aluminum flashing material used on the outer cover, bending the corners takes a little patience and practice.

✔ **Cost:** Using scrap wood (if you can find some) would keep material costs of this design minimal, but even if you purchase the recommended lumber, hardware, and fasteners, you can likely build this hive for less than $160 (a little less if you use knotty pine lumber).

Materials List

The following table lists what you'll use to build your Langstroth hive. In most cases, you can make substitutions as needed or desired. (See Chapter 3 for help choosing alternate materials.)

Lumber	Hardware	Fasteners
3 8' lengths of 1" x 12" clear pine lumber	Roll of 20" wide aluminum flashing (usually comes in a 10' length)	85 6 x 1⅝" deck screws, galvanized, #2 Phillips drive, flat-head with coarse thread and sharp point
1 2' x 4' sheet of ¾" thick exterior plywood	Optional: weatherproof wood glue	250 6d x 2" galvanized nails
1 2' x 4' sheet of ¼" thick lauan plywood	Optional: a gallon of latex or oil exterior paint (white or any light color), exterior polyurethane, or marine varnish	25 #8 x ½" lath screws, galvanized, #2 Phillips drive, flat-head with sharp point

Here are some notes about the materials for your Langstroth hive:

✔ I like using clear pine; it's not too expensive as lumber goes. Alternatively, knotty pine is even less expensive, but you may have to order extra material if a random knot interferes with the joinery of parts. You can also use different kinds of wood for your Langstroth hive. Cedar and cypress make beautiful hives, for example, and you can really get fancy with a cherry hive. It's up to you.

✔ Depending on where you buy it, plywood sometimes comes as 23/32 inch (rather than 3/4 inch). No worries: The difference is minimal, and either way, the plywood will fit just fine.

✔ I've included a few more fasteners than you'll use because, if you're like me, you'll lose or bend a few along the way. It's better to have a few extras on hand and save yourself another trip to the hardware store.

Cut List

The following sections break down the Langstroth hive into its individual components and provide instructions on how to cut and put together those components. (I provide cut lists for both ten- and eight-frame versions of this hive.)

Lumber in a store is identified by its *nominal* size, which is its rough dimension before it's trimmed and sanded to its finished size at the lumber mill. The actual finished dimensions are always slightly different from the nominal dimensions. For example, what a lumberyard calls *1 inch x 6 inch lumber* is in fact 3/4 inch x 5½ inch, and *1 inch x 12 inch lumber* is in fact 3/4 inch x 11½ inch. In the following sections, each Material column lists nominal dimensions, and each Dimensions column lists the actual, final measurements.

If finger joints are just a little too tricky for you, you can use rabbet joints for the deep hive bodies and the medium super. A rabbet joint may not be as strong as a finger joint, but with some weatherproof wood glue plus the nails, it should do the trick. If you choose the simpler rabbet option, you need to cut a 3/4 inch wide by 3/8 inch deep rabbet along each corner of the short sides of the hive bodies and the short sides of the medium super. Also, you need to cut the long sides of the hive bodies and mediums to 19⅛ inch when using the rabbet option (rather than the 19⅞ inch length when using the finger joint option). Flip to Chapter 4 for more about cutting rabbet joints.

This hive design makes use of rabbet cuts, dado cuts, and finger joints. See Chapter 4 for detailed instructions on making all these cuts.

Bottom board (ten-frame version)

Quantity	Material	Dimensions	Notes
2	1" x 12" clear pine	22" x 1⅛" x ¾"	These are the side rails. Dado ¾" wide by ⅜" deep along the entire length of each side rail, and rabbet one of the rear corners of each side rail ¾" wide by ⅜" deep.
1	1" x 12" clear pine	15½" x 1⅛" x ¾"	This is the rear rail. Dado ¾" wide by ⅜" deep along the entire length.
1	1" x 12" clear pine	14⅝" x ¾" x ¾"	This is the entrance reducer. Cut two notches on two different sides of the entrance reducer (one side ⅜" high by ¾" wide, and the other side ⅜" high by 4" wide).
1	¾" exterior plywood	21⅝" x 15½" x ¾"	This is the floor.

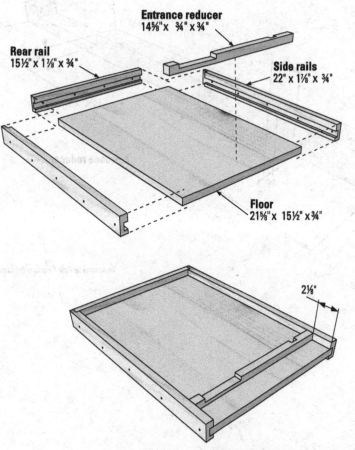

Entrance reducer
14⅝"x ¾"x ¾"

Rear rail
15½" x 1⅛" x ¾"

Side rails
22" x 1⅛" x ¾"

Floor
21⅝"x 15½" x ¾"

2⅛"

Illustration by Felix Freudzon, Freudzon Design

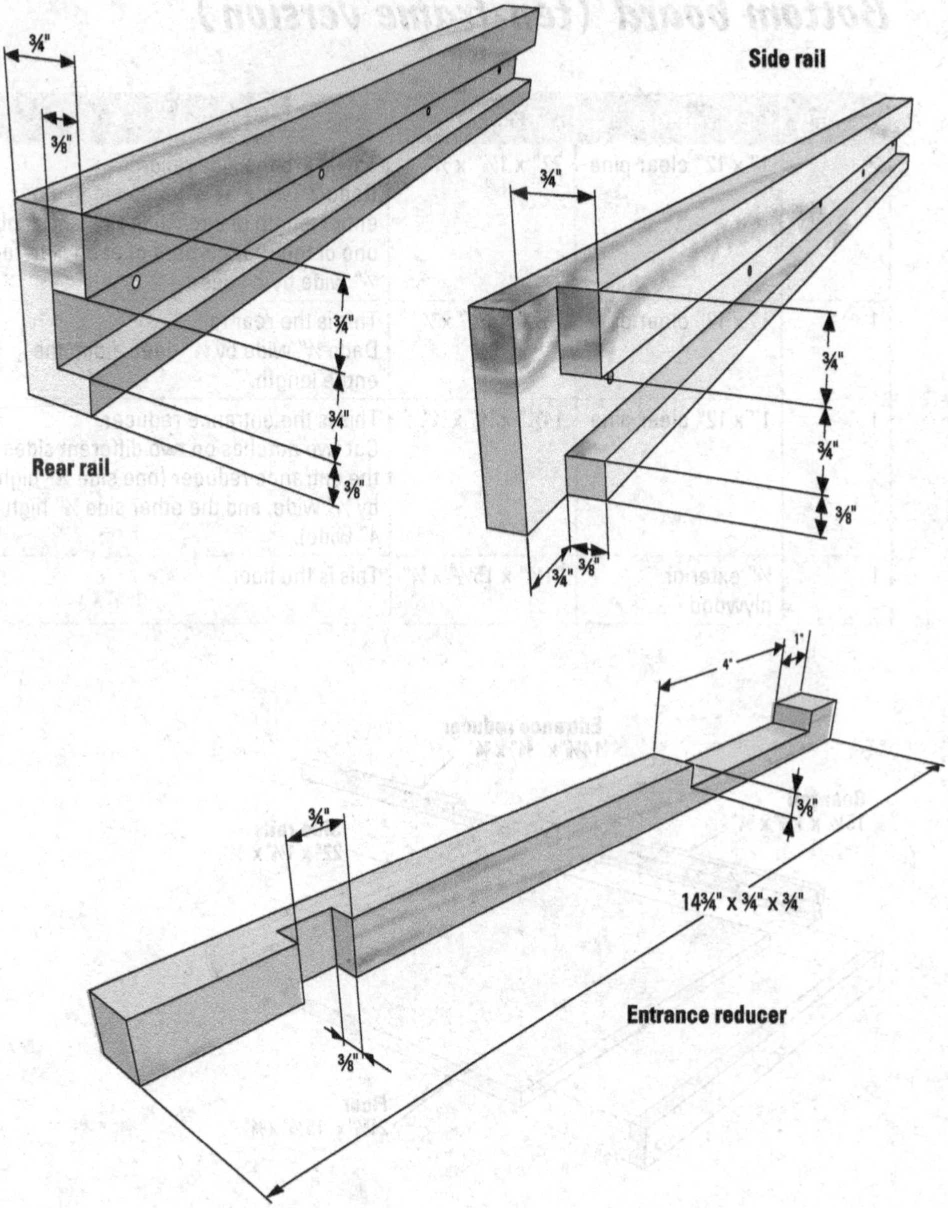
Side rail
3/4"
3/8"
3/4"
3/4"
3/4"
3/8"
Rear rail
3/4"
3/4"
3/8"
3/4"
3/8"
4"
1"
3/8"
3/4"
14¾" x ¾" x ¾"
Entrance reducer
3/8"

Illustration by Felix Freudzon, Freudzon Design

Bottom board (eight-frame version)

Quantity	Material	Dimensions	Notes
2	1" x 12" clear pine	22" x 1⅞" x ¾"	These are the side rails. Dado ¾" wide by ⅜" deep along the entire length of each side rail, and rabbet one of the rear corners of each side rail ¾" wide by ⅜" deep.
1	1" x 12" clear pine	13" x 1⅞" x ¾"	This is the rear rail. Dado ¾" wide by ⅜" deep along the entire length.
1	1" x 12" clear pine	12⅛" x ¾" x ¾"	This is the entrance reducer. Cut two notches on two different sides of the entrance reducer (one side ⅜" high by ¾" wide, and the other side ⅜" high by 4" wide).
1	¾" exterior plywood	21⅝" x 13" x ¾"	This is the floor.

Deep hive bodies (ten-frame version)

Quantity	Material	Dimensions	Notes
4	1" x 12" clear pine	19⅞" x 9⅝" x ¾"	These are the long sides. For the ¾" finger joints, start your first cut ¾" from the bottom. Note that the top finger is ⅝" (not ¾") and is trimmed to ⅜" long.
4	1" x 12" clear pine	16¼" x 9⅝" x ¾"	These are the short sides. For the ¾" finger joints, start your first cut at the bottom. Note that the top finger is 1⅜" (not ¾"). Rabbet a cut ⅝" wide by ⅜" deep along the entire inside top length.
4	1" x 12" clear pine	16¼" x 1⅞" x ¾"	These are the hand rails.

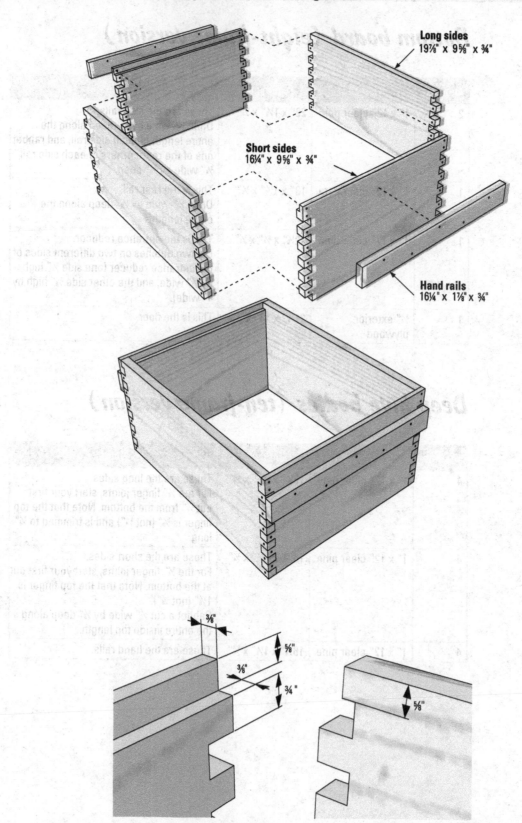

Long sides
19⅞" x 9⅝" x ¾"

Short sides
16¼" x 9⅝" x ¾"

Hand rails
16¼" x 1⅞" x ¾"

⅜"

⅜"

¾"

⅝"

⅝"

Illustration by Felix Freudzon, Freudzon Design

Deep hive bodies (eight-frame version)

Quantity	Material	Dimensions	Notes
4	1" x 12" clear pine	19⅞" x 9⅝" x ¾"	These are the long sides. For the ¾" finger joints, start your first cut ¾" from the bottom. Note that the top finger is ⅝" (not ¾") and is trimmed to ⅜" long.
4	1" x 12" clear pine	13¾" x 9⅝" x ¾"	These are the short sides. For the ¾" finger joints, start your first cut at the bottom. Note that the top finger is 1⅜" (not ¾"). Rabbet a cut ⅝" wide by ⅜" deep along the entire inside top length.
4	1" x 12" clear pine	13¾" x 1⅞" x ¾"	These are the hand rails.

Medium super (ten-frame version)

Quantity	Material	Dimensions	Notes
2	1" x 12" clear pine	19⅞" x 6⅝" x ¾"	These are the long sides. For the ¾" finger joints, start your first cut ¾" from the bottom. Note that the top finger is ⅝" (not ¾") and is trimmed to ⅜" long.
2	1" x 12" clear pine	16¼" x 6⅝" x ¾"	These are the short sides. For the ¾" finger joints, start your first cut at the bottom. Note that the top finger is 1⅜" (not ¾"). Rabbet a groove ⅝" wide by ⅜" deep along the entire inside top length.
2	1" x 12" clear pine	16¼" x 1⅞" x ¾"	These are the hand rails.

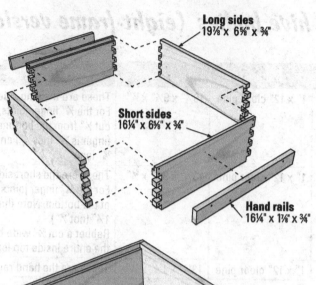

Long sides
19⅞" x 6⅝" x ¾"

Short sides
16¼" x 6⅝" x ¾"

Hand rails
16¼" x 1⅛" x ¾"

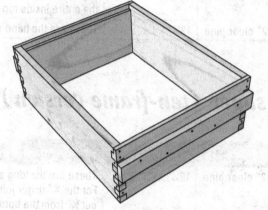

Illustration by Felix Freudzon, Freudzon Design

Medium super (eight-frame version)

Quantity	Material	Dimensions	Notes
2	1" x 12" clear pine	19⅞" x 6⅝" x ¾"	These are the long sides. For the ¾" finger joints, start your first cut ¾" from the bottom. Note that the top finger is ⅝" (not ¾") and is trimmed to ⅝" long.
2	1" x 12" clear pine	13¾" x 6⅝" x ¾"	These are the short sides. For the ¾" finger joints, start your first cut at the bottom. Note that the top finger is 1⅜" (not ¾"). Rabbet a groove ⅝" wide by ⅜" deep along the entire inside top length.
2	1" x 12" clear pine	13¾" x 1⅛" x ¾"	These are the hand rails.

Inner hive cover (ten-frame version)

Quantity	Material	Dimensions	Notes
2	1" x 12" clear pine	19⅛" x ¾" x ¾"	These are the long rails. Dado ¼" wide by ⅜" deep along entire length, ⅛" from edge.
2	1" x 12" clear pine	15½" x ¾" x ¾"	These are the short rails. Dado ¼" wide by ⅜" deep along entire length, ⅛" from edge. Option: Cut a ¾" wide by ¼" deep ventilation notch on the center point of *one* short rail (on the thick side of the rail).
1	¼" lauan plywood	19⅛" x 15⁷⁄₁₆" x ¼"	This is the top. Drill a 1" round ventilation hole in the center of the plywood.

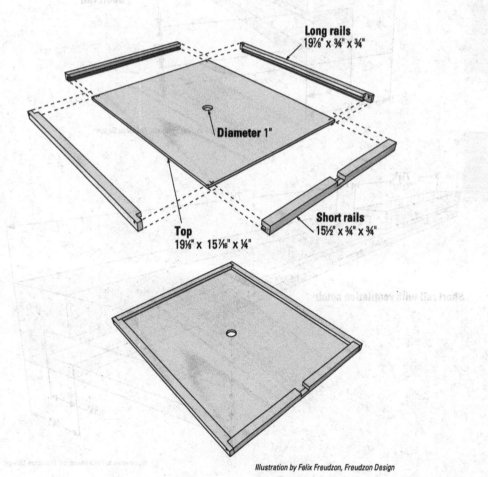

Long rails
19⅛" x ¾" x ¾"

Diameter 1"

Top
19⅛" x 15⁷⁄₁₆" x ¼"

Short rails
15½" x ¾" x ¾"

Illustration by Felix Freudzon, Freudzon Design

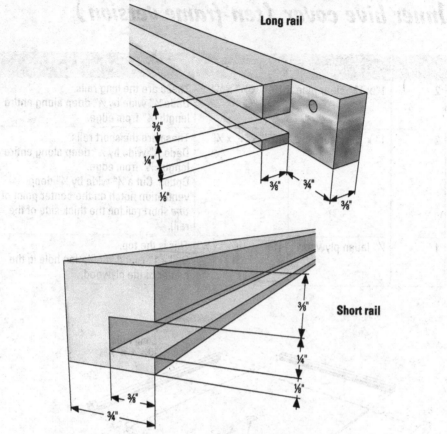

Long rail

Short rail

⅜"

¼"

⅛"

⅜"

¾"

Illustration by Felix Freudzon, Freudzon Design

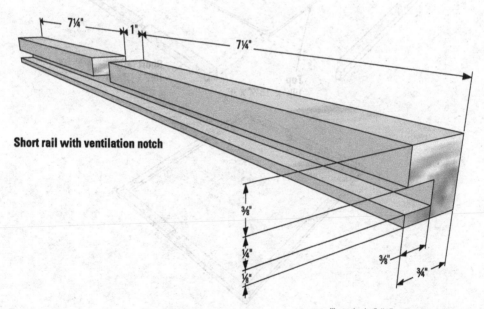

Short rail with ventilation notch

7¼"

1"

7¼"

⅜"

¼"

⅛"

⅜"

¾"

Illustration by Felix Freudzon, Freudzon Design

Inner hive cover (eight-frame version)

Quantity	Material	Dimensions	Notes
2	1" x 12" clear pine	19⅛" x ¾" x ¾"	These are the long rails. Dado ¼" wide by ⅜" deep along entire length, ⅛" from edge.
2	1" x 12" clear pine	13" x ¾" x ¾"	These are the short rails. Dado ¼" wide by ⅜" deep along entire length, ⅛" from edge. Option: Cut a ¾" wide by ¼" deep ventilation notch on the center point of *one* short rail (on the thick side of the rail).
1	¼" lauan plywood	19⅛" x 13" x ¼"	This is the top. Drill a 1" round ventilation hole in the center of the plywood.

Outer hive cover (ten-frame version)

Quantity	Material	Dimensions	Notes
2	1" x 12" clear pine	21¼" x 2¼" x ¾"	These are the long rails. Rabbet ¾" wide by ⅜" deep along entire length of top edge.
2	1" x 12" clear pine	18¼" x 2¼" x ¾"	These are the short rails. Rabbet ¾" wide by ⅜" deep along entire length of top edge. Also rabbet ¾" wide by ⅜" deep along both ends of the boards.
1	¾" exterior plywood	21¼" x 17½" x ¾"	This is the top.
1	20" aluminum flashing	23¾" x 20"	This is the protective metal roof. Wrap the flashing over the top of the assembled outer cover. There will be a ⅞" lip folded over the edges of the top. Fold the corners to avoid sharp edges.

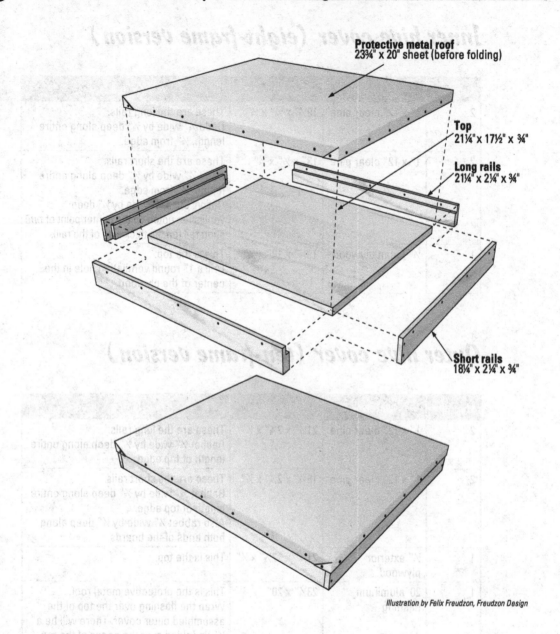

Protective metal roof
23¾" x 20" sheet (before folding)

Top
21¼" x 17½" x ¾"

Long rails
21¼" x 2¼" x ¾"

Short rails
18¼" x 2¼" x ¾"

Illustration by Felix Freudzon, Freudzon Design

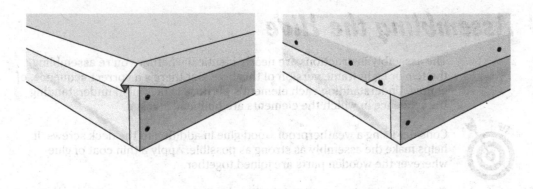

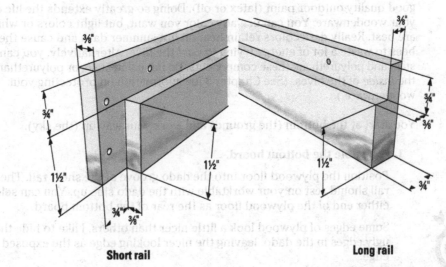

Short rail

Long rail

Illustration by Felix Freudzon, Freudzon Design

Outer hive cover (eight-frame version)

Quantity	Material	Dimensions	Notes
2	1" x 12" clear pine	21¼" x 2¼" x ¾"	These are the long rails. Rabbet ¾" wide by ⅜" deep along entire length of top edge.
2	1" x 12" clear pine	15¾" x 2¼" x ¾"	These are the short rails. Rabbet ¾" wide by ⅜" deep along entire length of top edge. Also rabbet ¾" wide by ⅜" deep along both ends of the boards.
1	¾" exterior plywood	21¼" x 15" x ¾"	This is the top.
1	20" aluminum flashing	17¾" x 22"	This is the protective metal roof. Wrap the flashing over the top of the assembled outer cover. There will be a 1" lip folded over the edges of the top. Fold the corners to avoid sharp edges.

Assembling the Hive

The assembly instructions are nearly identical whether you're assembling the ten- or eight-frame version of this hive. But there's a correct sequence, of course. Understanding each element's purpose is helpful in understanding the sequence in which the elements are built and stacked.

Consider using a weatherproof wood glue in addition to the deck screws. It helps make the assembly as strong as possible. Apply a thin coat of glue wherever the wooden parts are joined together.

Optional: Paint the outside of the hive bodies, super, and outer cover with a good quality outdoor paint (latex or oil). Doing so greatly extends the life of your woodenware. You can use any color you want, but light colors or white are best. Really dark colors retain heat on hot summer days and cause the bees to waste a lot of energy trying to cool the hive. Alternatively, you can stain and polyurethane these components. Do not paint, stain, or polyurethane the *inside* of the hives. (See Chapter 3 for information on protecting your woodenware.)

You start at the bottom (the ground) and work your way up (the sky).

1. **Assemble the bottom board.**

   Position the plywood floor into the dado groove of the short rail. The rail should rest on your worktable with the dado side up. You can select either end of the plywood floor as the rear of the bottom board.

   Some edges of plywood look a little nicer than others. I like to hide the ugly edges in the dado, leaving the nicer looking edge as the exposed edge.

   Place the long rails on both sides of the plywood floor, inserting the plywood into the dados.

   Be certain that the dado faces the same way in all rails (the dado isn't centered along the rail). Otherwise, you'll have a seriously lopsided bottom board!

   Check the alignment and fit of all the rails with the floor, and then place one of the #6 x 1⅝ inch galvanized deck screws halfway into the center of each of the three rails (the screws go through the rails and into the edges of the plywood) with a drill. Don't screw them in all the way yet. First make sure that everything fits properly; you have no room for adjustment after all the screws are in! When the fit looks good, use four additional deck screws spaced evenly (by eye) along each long rail, and three additional screws spaced evenly (by eye) along the short rail.

   The screws will go in easier if you first drill a 7⁄64" hole in each spot you plan to place a screw. The pre-drilling also helps prevent the wood from splitting.

The entrance reducer remains loose, and you place it in the entrance of the hive to control ventilation and prevent *robbing,* that dreaded occurrence where aggressive invaders from another colony steal all the honey from your hive. The entrance reducer is typically not used year-round. For more information on using an entrance reducer, see Chapter 2.

2. Assemble the deep hive bodies.

You'll be assembling two deep hive bodies. Each goes together in the identical manner. First, assemble the two long sides and the two short sides by tapping the finger joints together with a rubber mallet. You're essentially building a box. If the fit is too snug, use 60 grit sandpaper to remove some wood material from the offending fingers.

Use a carpenter's square to make sure the box stays square as you assemble the hive body because you won't have an opportunity for correction after all the nails are in place! (See Chapter 4 for more information on keeping things square.)

When the "dry" fit looks good and everything is squared up, begin to lock the finger joints in place by nailing a 6d x 2 inch galvanized nail into one of the center fingers on each of the hive's four corners. Hammer the nail in only *halfway* to make sure everything remains square and fits properly, make sure everything looks okay, and then hammer the remaining nails all the way in. You use one 6d x 2 inch nail in each of the fingers and an extra one for strength in the wider top finger (see the later figure).

Consider using a ⁷⁄₆₄" bit to drill a hole in the center of each finger joint (these holes make it easier for the nails to go in and prevent the wood from splitting).

Now use the deck screws to attach the two hand rails to the short sides of the hive body. Position the top edge of the hand rails 2 inches down from the top edge of the hive body. Use five screws per hand rail, spaced and staggered as shown in the figure (mathematical precision isn't necessary). The staggering prevents the wood from splitting.

In place of using wooden hand rails, you can attach flush-mounted, galvanized (or stainless steel) handles to the hive body. They provide you with a much more authoritative grip, and they look kind of cool. You can find these handles in hardware stores or marine supply stores.

Check all sides to make certain that all the nails and screws are in place. All good? You're done making one deep hive body. Assemble the second one in the same manner.

3. Assemble the medium super.

Use a ⁷⁄₆₄ inch bit to drill a hole in the center of each finger joint.

Assemble the two long sides and the two short sides by tapping the finger joints together with a rubber mallet. Again, you're essentially building a box. If the fit is too snug, use 60 grit sandpaper to remove some wood material from the offending fingers.

Use a carpenter's square to make sure the box stays square as you assemble the hive body because you won't have an opportunity for correction after all the screws are in place!

When the "dry" fit looks good and everything is squared up, begin to lock the finger joints in place by nailing a 6d x 2 inch galvanized nail into one of the center fingers on each of the super's four corners. Hammer the nail in only *halfway* to make sure everything is square and fits properly, make sure everything looks okay, and then hammer the nails all the way in. You use one 6d x 2 inch nail in each of the fingers, plus an extra one in the wide top finger (see the figure).

Now use the deck screws to attach the two hand rails to the short sides of the hive body. Position the top edge of the hand rails 2 inches down from the top edge of the hive body. Use five screws per hand rail, spaced and staggered as shown in the figure (mathematical precision isn't necessary).

In place of using wooden hand rails, you can attach flush-mounted, galvanized (or stainless steel) handles to the super.

Check all sides to make certain that all the nails are in place. Congratulations! You're done making the medium super!

4. Assemble the inner cover.

Position the plywood cover insert into the dado of the long rails and into the dado of the short rails. This is kind of like putting a picture frame together.

Be certain that *all* rails have the *thick* or *thin* lip of the groove facing the same way. Otherwise, you'll have a seriously lopsided inner cover!

Check the alignment and fit, and insert a deck screw halfway into each of the four corners. Don't put the screws all the way in yet. First make sure everything is square and fits properly. When all looks good, you can screw them in all the way. Keep in mind that if you cut the plywood insert square, it will essentially square up the frame.

Don't paint the inner cover. Leave it natural and unfinished, as with all the internal parts of any beehive.

Note: You should position the inner cover on the hive body with the flat side down and with the cutout notch (bee ventilation/entrance) facing up and at the front of the hive.

5. Assemble the outer cover.

Start with one long rail. Insert the plywood into the rabbeted groove. Repeat this step on the opposite side with the second long rail.

Fit both of the two short rails onto the plywood board. The rails form a frame surrounding the plywood board. If the plywood was cut square, it helps square up the entire assembly.

When assembling the outer cover, have a "stop" on your worktable that you can push against while inserting screws. A short 2x4 of lumber clamped to the table serves as a good stop to work against.

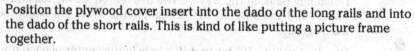

Place one end of the outer cover flat on the worktable against the stop. Using deck screws, insert two screws into each corner of the short rails. Reverse the entire cover end to end, and screw the other corners of the short rails in a similar manner. Make certain the entire assembly remains snug and tight as you do this.

Now use the deck screws to secure the plywood insert to the assembly. Drive the screws through the rails and into the edges of the plywood board. Five evenly spaced screws along each long rail and four along the short rails should do the trick.

Center the aluminum flashing evenly on the top of the outer cover, and bend the flashing over the edges of the rail/frame. This creates a ⅛ inch lip all around the top edge. Do this to all four sides. Bend and fold the corners (like you're making the corners of a bed). The flashing is thin and fairly easy to work with. Use a rubber mallet to coax the corners flush and flat.

The edges of aluminum flashing are very sharp. Use caution when handling flashing to avoid cutting yourself, and consider using work gloves.

Affix the folded edges of the flashing to the outer cover using the #8 x ½ inch lath screws. I use six lath screws evenly spaced by eye per short side and six along each long side, plus an extra screw to secure each folded corner.

6. **Stack all the pieces together to create the hive.**

    Place the bottom board on level ground. The bottom board is the floor of the hive. It keeps the colony off the damp ground and provides for the entrance of the hive (where the bees fly in and out).

    Consider using an elevated hive stand to raise the hive farther off the ground, to make it more accessible to you, and to improve ventilation. (See Chapter 13 for instructions on building an elevated hive stand; your back will thank you!)

    The two deep hive bodies go on top of the bottom board. The bees raise baby bees and store food in these boxes. The bees tend to use the lower deep for raising brood and the upper deep for storing food. A nursery and a pantry! Place deep frames with foundation into each deep hive body (either ten or eight frames, depending on which version of the hive you're building).

    Stack the medium super on top of the deep hive bodies. This is where the bees store extra honey. That's the honey you harvest for yourself. One medium super holds about 35 pounds of honey. Place medium frames with foundation into the medium honey super (either ten or eight frames, depending on which version of the hive you're building).

    Some beekeepers use a queen excluder placed between the top deep and the medium honey super above. As the name implies, this gizmo prevents the queen from entering the honey super, where she might start laying eggs. A queen laying eggs in the super encourages the other bees to bring pollen into the super, spoiling the clarity of the honey.

When your medium honey super is about half-filled with capped honey, it's time to build another medium super and more medium frames with foundation. If you're lucky and the honey flow is heavy, you may ultimately stack three, four, or more medium supers on your hive. That's a honey bonanza!

You place the inner cover on top of the medium super. The deeper ledge faces up. If you choose to cut a ventilation notch in the inner cover, it faces toward the front (entrance) of the hive.

You remove the inner cover if you use a hivetop feeder on your hive. (For instructions on making a hive-top feeder for this hive, see Chapter 15.)

The outer cover is the roof of your hive, providing protection from the elements. The Langstroth hive uses a *telescoping cover* design, meaning the cover fits on and over the hive (like a hat). Simply stack the outer cover on top of your inner cover, and that's it. Your Langstroth hive is ready for the bees to settle in!

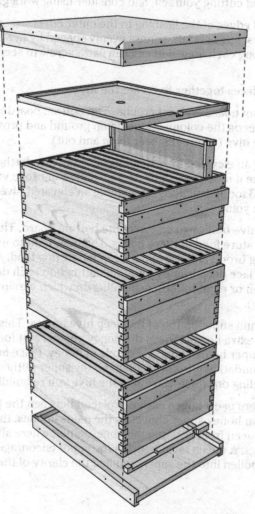

Illustration by Felix Freudzon, Freudzon Design

Part III
Sweet Beehive Accessories

The 5th Wave   By Rich Tennant

"I try to see to it that it's not all work for my bees."

In this part . . .

Beehives are only part of this building adventure. In this part, I include plans for making some really cool and practical beekeeping accessories: a frame jig, a double screened inner cover, an elevated hive stand, a screened bottom board, a hive-top feeder, a solar wax melter, and made-from-scratch Langstroth-style frames. Each chapter includes a shopping list of the materials you need, an illustrated cut list that shows you how to trim your lumber, and detailed assembly instructions to help you put it all together.

Chapter 11

The Frame Jig

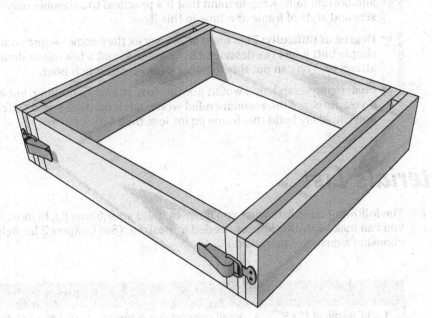

Illustration by Felix Freudzon, Freudzon Design

Whether you make your own frames from scratch (see Chapter 17) or purchase frames from a beekeeping supplier, you still have to assemble, glue, and nail the parts together. At best it's a monotonous chore. But you can make the task considerably easier (and certainly faster) by using a nifty little gadget called a *frame jig*. It's hard to find this item in stores, but you can make your very own, allowing you to spend less time assembling frames and more time with your bees.

The frame jig you build in this chapter holds all the parts to assemble and nail ten frames at a time (either deep, medium, or shallow). Assembling frames is so much easier when you don't have to juggle and nail frame parts one bit at a time. It's like having your own assembly line! You're going to love having this frame jig.

Frame jigs make terrific gifts for other beekeepers. Consider making a presentation-grade model for your favorite beekeeper using fancy wood — now you're talking!

Vital Stats

- ✔ **Size:** 19⅝ inches x 16⅞ inches x 4 inches.

- ✔ **Capacity:** The frame jig holds up to ten shallow, medium, or deep Langstroth-style frames (see Chapter 17 for details on assembling frames).

- ✔ **Universality:** This jig best accommodates a Langstroth-style frame, although it also works with some other frame types that use top, side, and bottom rails. Keep in mind that it's practical to assemble only one size and style of frame at a time in this jig.

- ✔ **Degree of difficulty:** This build is as easy as they come — just some simple butt joints (as described in Chapter 4) and a few minor details to attend to. You can put this together during your lunch hour.

- ✔ **Cost:** Using scrap wood would put the cost at next to nothing, but even if you purchase the recommended wood, latch hardware, and fasteners, you can likely build this frame jig for less than $20.

Materials List

The following table lists what you'll use to build your frame jig. In most cases, you can make substitutions as needed or desired. (See Chapter 3 for help choosing alternative materials.)

Lumber	Hardware	Fasteners
1 10' length of 1" x 5" knotty pine lumber	4 small compression draw (toggle) latches, catch strikes, and screws. This set of hardware typically comes in a blister pack with fasteners and is available at most hardware stores. The ones I use are about 2½" long when assembled.	10 #6 x 1⅜" deck screws, galvanized, #2 Phillips drive, flat-head with coarse thread and sharp point
	Optional: weatherproof wood glue	10 ⁵⁄₃₂" x 1⅛" flat-head, diamond-point wire nails

I've tossed in a few more screws and nails than you'll actually use. If you're like me, you'll lose a few along the way. It's better to have a few extras on hand and save another trip to the hardware store.

Cut List

This section breaks down the frame jig into its individual components and provides instructions on how to cut those components.

Lumber in a store is identified by its *nominal* size, which is its rough dimension before it's trimmed and sanded to its finished size at the lumber mill. The actual finished dimensions are always slightly different from the nominal dimensions. For example, what a lumberyard calls *1 inch x 5 inch lumber* is in fact ¾ inch x 4½ inch. The Material column in the following table lists nominal dimensions, and the Dimensions column lists the actual, final measurements.

Quantity	Material	Dimensions	Notes
4	1" x 5" knotty pine	4" x 1⅜" x ½"	These are spacer cleats for the frame retaining panels.
2	1" x 5" knotty pine	16⅞" x 4" x ¾"	These are the retaining panels that hold the frames in place.
2	1" x 5" knotty pine	16⅞" x 4" x ¾"	These are the long sides.
2	1" x 5" knotty pine	15⅜" x 4" x ¾"	These are the short sides.

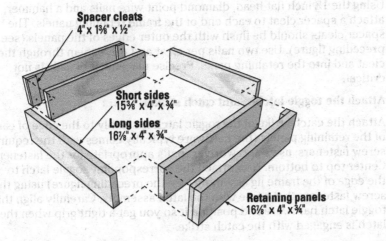

Spacer cleats
4" x 1⅜" x ½"

Short sides
15⅜" x 4" x ¾"

Long sides
16⅞" x 4" x ¾"

Retaining panels
16⅞" x 4" x ¾"

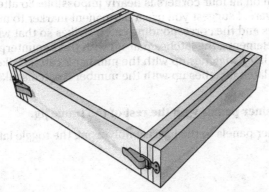

Illustration by Felix Freudzon, Freudzon Design

Assembling the Frame Jig

Clear off your workbench: It's time to assemble all the pieces of your jig.

1. **Attach the long sides to the short sides.**

   Align the ends of the long sides with the edges of the short sides. Use the deck screws and your power drill with a #2 Phillips head bit to attach the long sides to the edges of the short sides. These are butt joints (see Chapter 4). Use two screws per corner, spaced as shown in the preceding figure.

   The screws will go in easier if you first drill a ⁷⁄₆₄-inch hole in each spot you plan to place a screw. The pre-drilling also helps prevent the wood from splitting.

   Throughout this process, consider using wood glue in addition to the screws. It helps make the jig as strong as possible. Apply a thin coat of glue wherever the wooden parts are permanently joined together.

2. **Nail the spacer cleats onto the frame retaining panels.**

   Using the 1⅛ inch flat-head, diamond-point wire nails and a hammer, attach a spacer cleat to each end of the frame retaining panels. The spacer cleats should be flush with the outer edges of the panels (see the preceding figure). Use two nails per cleat and drive them through the cleat and into the retaining panel. Precise placement of nails is not critical.

3. **Attach the toggle latches and catch strikes.**

   Attach the catch strike of the toggle latch assembly to the edge of each of the retaining panels (the hardware typically comes with the required screw fasteners; use a screwdriver that's appropriate for the fasteners). Center top to bottom. Now attach the corresponding toggle latch to the edge of the frame jig assembly (see the preceding figure) using the screw fasteners that come with the latch assembly. Carefully align this toggle latch hardware and position it so you get a tight grip when the latch is engaged with the catch strike.

   Aligning the placement of the four toggle latches in a mathematically identical position on all four corners is nearly impossible. So after you attach the hardware, I suggest you use a permanent marker to number the toggle latches and the corresponding catch strikes so that when you reassemble the elements, the latches pair up with their counterpart (the number 2 toggle latch matches up with the number 2 catch strike, the number 3 toggle latch matches up with the number 3 catch strike, and so on).

4. **Connect the retainer panels with the rest of the frame jig.**

   Fasten the retainer panels to the jig assembly using the toggle latch hardware.

You're now ready to become a frame-making machine! Just follow these steps to use your jig:

1. **Insert 10 side bars into *each* of the channels created by the frame retaining bars (20 total).**

   The wide end of the side bar should be facing up.

2. **Snap the top bars into the slot at the wide end of each pair of side bars and secure the top bars in place using two nails at each end.**

   In addition to nailing, you can also use a little wood glue.

3. **Turn the whole jig over (upside down) so that the top bars are now resting on the work surface.**

4. **Insert the bottom bars into the slot at the narrow end of the side bars and secure the bottom bars using a nail at each end.**

5. **Open the toggle latches and remove the retainer panels so you can slide the frames free of the jig.**

Now isn't that easier than assembling frames one at a time?

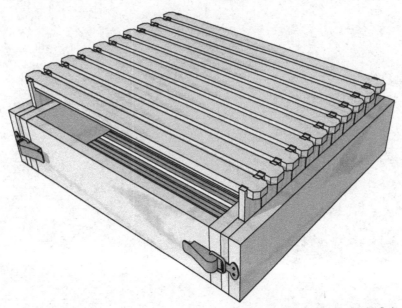

Illustration by Felix Freudzon, Freudzon Design

Chapter 12
The Double Screened Inner Cover

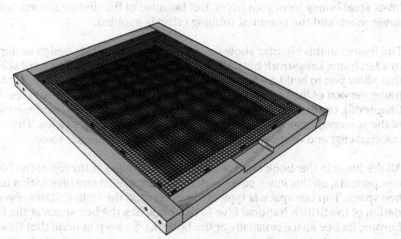

Illustration by Felix Freudzon, Freudzon Design

More and more backyard beekeepers are using screened inner covers (or screened crown boards) on their hives. They use them in place of the standard inner covers and crown boards. Why the growing interest? Improved ventilation. Poor ventilation is one of the leading causes of stress on a colony. Using a screened inner cover provides the ultimate in ventilation. Some beekeepers leave the screened inner cover on the colony all year long, even in cold weather climates.

You can also use a screened inner cover to move a colony on a hot day. Just remove the outer cover and snugly strap the screened inner cover in place. Make sure to position it on the hive such that the entry notch is closed. You'll have ample cooling and ventilation for the colony while you move it to a different location.

The design in this chapter was developed by friend and fellow beekeeper Michael Lund. Michael is brilliant at solving beekeeping problems, particularly when it comes to designing bee equipment. This design is extra heavy-duty compared to lightweight commercial designs that tend to twist and become damaged in use.

Also, the double screen in this design has a specific purpose. Placing the cover on a hive one way provides a ⅜" upper entrance for the bees; flipping it the other way results in a snug fit with no upper entrance option. If you flip the cover so that the entrance notch is closed, bees from other colonies can still get under the hive's outer cover. If there were only a single screen, they could make friends with your bees by extending their tongues and offering peace offerings of nectar. In due course, this seemingly friendly communication could turn out to be a bribe to gain entry to the hive and then escalate into a robbing situation (*robbing* is a nasty situation when bees from other hives steal honey from your hive). But because of the double screen, tongues never meet, and the potential robbing crisis is avoided.

The figures in this chapter show the measurements for this design as applied to a ten-frame Langstroth hive. However, I also include alternative cut sheets that allow you to build a screened inner cover that fits nicely on the eight-frame version of the Langstroth hive (see Chapter 10), the nuc hive (see Chapter 6), or the British National hive (see Chapter 9). Only the dimensions of the screened inner cover vary across these different hive types. The materials list and the assembly instructions are basically the same.

All the hives in this book are designed with bee space at the *top* of the hive components, so this inner cover is designed for use on any hive with a top bee space. Top bee space is typical for all hives in the United States. Even the design of the British National hive in this book has the bee space at the top (in Europe, its bee space is usually at the bottom). So keep in mind that this inner cover may not work well on a hive you purchase commercially that utilizes a *bottom* bee space.

Vital Stats

- ✔ **Size:** 19⅞ inches x 16¼ inches x 1⅛ inches (for the Langstroth ten-frame hive); 19⅞ inches x 13¾ inches x 1⅛ inches (for the Langstroth eight-frame hive); 19⅞ inches x 9 inches x 1⅛ inches (for the nuc hive); 18⅛ inches x 18⅛ inches x 1⅛ inches (for the British National hive).

- ✔ **Capacity:** You need to build one screened inner cover for each hive in your apiary.

- ✔ **Degree of difficulty:** This is likely one of the easiest projects in this book. It's no more involved than making a heavy-duty wooden frame and stapling hardware cloth to both sides.

- ✔ **Cost:** As with any of these hive parts, using scrap wood (got any left over from other projects in this book?) would keep material costs next to nothing. But even if you purchase the recommended lumber, fasteners, and hardware, you can likely build this screened inner cover for less than $15. What a steal!

Materials List

Here's what you'll use to build your screened inner cover. In most cases, you can make substitutions as needed or desired. (See Chapter 3 for help choosing alternative materials.)

Lumber	Hardware	Fasteners
1 6' length of ¾" x 3" clear or knotty pine lumber	24" of ⅛" hardware cloth. (#8 typically comes in 3' wide by 10' long rolls, but some beekeeping supply stores sell it by the foot.)	10 #6 x 2½" deck screws, galvanized, #2 Phillips drive, flat-head with coarse thread and sharp point
	Optional: weatherproof wood glue	40 ⅜" staples for use in a heavy-duty staple gun

Here are a few details about the materials for your double screened inner cover:

✔ I use pine for my screened inner covers because it's easy to find and not terribly expensive. You can use clear pine or the less expensive knotty pine — it's up to you. I prefer clear pine because it's a little easier to work with (tight grain and no knots).

✔ I've tossed in some extra fasteners in case you bend or lose a few along the way. It's better to have a few extras on hand and save yourself an extra trip to the hardware store.

✔ You may want a screened inner cover for each hive in your apiary. So if you have a couple of hives, you need to double the materials list to build a couple of inner covers.

Cut List

The following sections provide instructions on how to cut the individual components for each type of double screened inner cover: the ten-frame Langstroth hive, the eight-frame Langstroth hive, the nuc hive, and the British National hive.

Lumber in a store is identified by its *nominal* size, which is its rough dimension before it's trimmed and sanded to its finished size at the lumber mill. The actual finished dimensions are always slightly different from the nominal dimensions. For example, what a lumberyard calls 5⁄4 inch by 3 inch lumber is in fact 1⅛ inch by 2½ inch. In the following tables, each Material column lists nominal dimensions, and each Dimensions column lists the actual, final measurements.

This design makes use of rabbet cuts. See Chapter 4 for detailed instructions on making this type of cut.

Double screened inner cover for the ten-frame Langstroth hive

Quantity	Material	Dimensions	Notes
2	5⁄4" x 3" clear pine	19⅞" x 2¼" x 1⅛"	These are the long rails. Rabbet *both* sides of the inside edge 1" wide by ⅛" deep. This cut will accommodate the thickness of the hardware cloth. Cut a 2¼" x 1" notch at both ends of the long rails.
2	5⁄4" x 3" clear pine	13¾" x 2¼" x 1⅛"	These are the short rails. Rabbet *both* sides of the inside edge 1" wide by ⅛" deep. This cut will accommodate the thickness of the hardware cloth. Cut a 3" x 1¼" x ⅜" entrance notch centered along the edge of *one* of the short rails. This is most easily done with a chisel or a router if you have one.
2	#8 hardware cloth	17¼" x 13½"	This is the screened top and bottom.

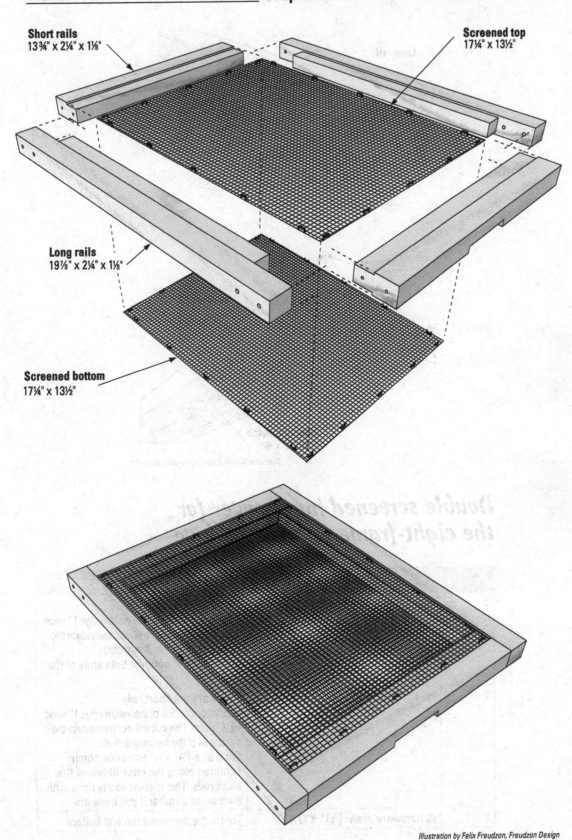

Short rails
13 ¾" x 2¼" x 1⅛"

Screened top
17¼" x 13½"

Long rails
19⅞" x 2¼" x 1⅛"

Screened bottom
17¼" x 13½"

Illustration by Felix Freudzon, Freudzon Design

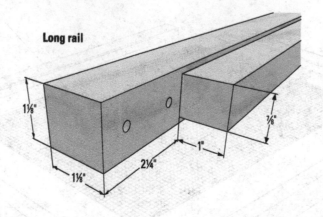

Long rail

$1\frac{1}{8}$"

$\frac{7}{8}$"

$1\frac{1}{8}$"

$2\frac{1}{4}$"

1"

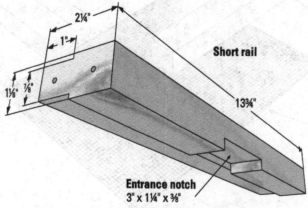

$2\frac{1}{4}$"

1"

Short rail

$1\frac{1}{8}$" $\frac{7}{8}$"

$13\frac{3}{4}$"

Entrance notch
3" x $1\frac{1}{4}$" x $\frac{3}{8}$"

Illustration by Felix Freudzon, Freudzon Design

Double screened inner cover for the eight-frame Langstroth hive

Quantity	Material	Dimensions	Notes
2	$\frac{5}{8}$" x 3" clear pine	$19\frac{7}{8}$" x $2\frac{1}{4}$" x $1\frac{1}{8}$"	These are the long rails. Rabbet *both* sides of the inside edge 1" wide by $\frac{1}{8}$" deep. This cut will accommodate the thickness of the hardware cloth. Cut a $2\frac{1}{4}$" x 1" notch at both ends of the long rails.
2	$\frac{5}{8}$" x 3" clear pine	$11\frac{1}{4}$" x $2\frac{1}{4}$" x $1\frac{1}{8}$"	These are the short rails. Rabbet *both* sides of the inside edge 1" wide by $\frac{1}{8}$" deep. This cut will accommodate the thickness of the hardware cloth. Cut a 3" x $1\frac{1}{4}$" x $\frac{3}{8}$" entrance notch centered along the edge of *one* of the short rails. This is most easily done with a chisel or a router if you have one.
2	#8 hardware cloth	11" x $17\frac{1}{4}$"	This is the screened top and bottom.

Double screened inner cover for the nuc hive

Quantity	Material	Dimensions	Notes
2	⁵⁄₄" x 3" clear pine	19⅞" x 2¼" x 1⅛"	These are the long rails. Rabbet *both* sides of the inside edge 1" wide by ⅛" deep. This cut will accommodate the thickness of the hardware cloth. Cut a 2¼" x 1" notch at both ends of the long rails.
2	⁵⁄₄" x 3" clear pine	6½" x 2¼" x 1⅛"	These are the short rails. Rabbet *both* sides of the inside edge 1" wide by ⅛" deep. This cut will accommodate the thickness of the hardware cloth. Cut a 3" x 1¼" x ⅜" entrance notch centered along the edge of *one* of the short rails. This is most easily done with a chisel or a router if you have one.
2	#8 hardware cloth	6¼" x 17¼"	This is the screened top and bottom.

Double screened inner cover for the British National hive

Quantity	Material	Dimensions	Notes
2	⁵⁄₄" x 3" clear pine	18⅛" x 2¼" x 1⅛"	These are the long rails. Rabbet *both* sides of the inside edge 1" wide by ⅛" deep. This cut will accommodate the thickness of the hardware cloth. Cut a 2¼" x 1" notch at both ends of the long rails.
2	⁵⁄₄" x 3" clear pine	15⅝" x 2¼" x 1⅛"	These are the short rails. Rabbet *both* sides of the inside edge 1" wide by ⅛" deep. This cut will accommodate the thickness of the hardware cloth. Cut a 3" x 1¼" x ⅜" entrance notch centered along the edge of *one* of the short rails. This is most easily done with a chisel or a router if you have one.
2	#8 hardware cloth	15½" x 15½"	This is the screened top and bottom.

Assembling the Inner Cover

It's time to work some woodworking magic and assemble these bits and pieces into a double screened inner cover. It's an easy build — just a frame with a couple of screened inserts.

1. **Assemble the frame.**

   The short rails align with the notches that were cut from the ends of the long rails. It's kind of like building a picture frame.

   Check alignment and fit, and insert one deck screw halfway into each corner of each long rail using your power drill with a #2 Phillips head bit. Place the screws through the long rails and into the short rails. Center each screw top to bottom and place it about ½ inch from the outside edge of the assembly. Don't put the screws all the way in quite yet. First make sure everything is square and fits properly (see Chapter 4 for more about using a square); then, when all looks good, you can screw them in all the way. You have no room for adjustment when all the screws are in! Now add one extra screw in each of the frame's corners. Place them centered top to bottom and spaced about 1 inch from the original four screws (each corner of the long rails will have two screws).

   Consider using a weatherproof wood glue in addition to the screws. It helps make the screened inner cover as strong as possible. Apply a thin coat of glue wherever the wooden parts are joined together.

2. **Attach the screening material and place the cover on your hive.**

   Attach the #8 hardware cloth to the "opening" of the frame (on both sides). The screening settles into the ⅛ inch deep rabbet cut you made. Use ⅜ inch staples spaced approximately 2 inches apart. Staples go around the entire perimeter of the screening. There should be no gaps where the bees could squeeze through.

   Don't paint the screened inner cover. Leave it natural and unfinished, as with all the internal parts of any beehive.

   Position the screened inner cover on the hive body with the notched entrance facing down and toward the front of the hive (when you want to create an upper entrance for your bees to travel in and out of the hive; see the image on the left in the following figure). Or flip it over to close the upper entrance to and from the hive (see the image on the right).

Access closed

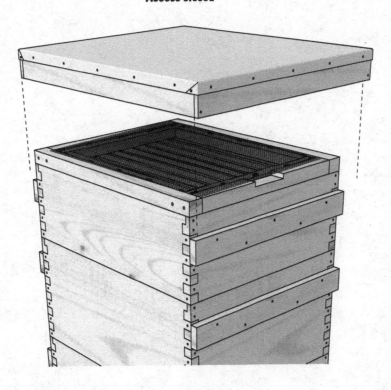

Access open

Illustration by Felix Freudzon, Freudzon Design

Chapter 13

The Elevated Hive Stand

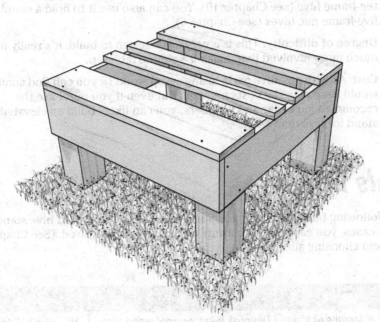

Illustration by Felix Freudzon, Freudzon Design

An elevated hive stand is exactly what it sounds like: an item you use to hold a beehive off the ground. I put all my hives on elevated hive stands like the one I show you how to build in this chapter. Doing so has several advantages:

✔ The hive stays well off the damp ground and provides better air circulation, which keeps the colony drier.

✔ You don't have to bend over to inspect the colony (your back will thank you).

✔ The hive's elevation helps deter skunks from snacking on your bees.

✔ If you choose to use an IPM screened bottom board, the (mostly) open slatted top of this particular hive stand design allows mites to drop to the ground, safely away from your bees (for instructions on building an IPM screened bottom board, see Chapter 14).

✔ The generous top not only accommodates your hive but also provides some extra surface area to place your smoker, tools, and the frames you remove for inspection.

Vital Stats

- ✔ **Size:** 24 inches x 24 inches x 13 inches.
- ✔ **Capacity:** This stand holds one hive, so you need to build one for each hive in your apiary.
- ✔ **Universality:** This design is ideal for the Warré hive (see Chapter 8), the British National hive (see Chapter 9), or the Langstroth eight- or ten-frame hive (see Chapter 10). You can also use it to hold a couple of five-frame nuc hives (see Chapter 6).
- ✔ **Degree of difficulty:** This is a very easy design to build. It's really not much more involved than making a simple tabletop.
- ✔ **Cost:** As with any hive parts, using scrap wood (if you can find some) would keep material costs minimal. But even if you purchase the recommended cedar and fasteners, you can likely build an elevated hive stand for less than $40.

Materials List

The following table lists what you'll use to build your elevated hive stand. In most cases, you can make substitutions as needed or desired. (See Chapter 3 for help choosing alternative materials.)

Lumber	Hardware	Fasteners
2 8' lengths of 1" x 6" cedar or knotty pine lumber	Optional: weatherproof wood glue	30 #6 x 2½" deck screws, galvanized, #2 Phillips drive, flat-head with coarse thread and sharp point
1 8' length of 4" x 4" cedar posts	Optional: a pint of exterior latex or oil paint, exterior polyurethane, or marine varnish	

Here are a few tips for purchasing materials for your elevated hive stand:

- ✔ I like to use cedar for my hive stands because it holds up very well outside. It's a little more expensive than pine, but it lasts much longer. Another suitable choice is cypress (if it's available in your neck of the woods).

✔ In the materials list, I've tossed in a few more deck screws than you'll need. It's better to have a few extras on hand and save another trip to the hardware store.

✔ You need a stand for each hive you build, so if you plan on having a couple of hives, you need to double everything in the preceding table to build a couple of stands.

Cut List

This section breaks down the elevated hive stand into its individual components and provides instructions on how to cut and assemble those components. (*Note:* This design calls for a rabbet cut. See Chapter 4 for detailed instructions on making this cut.)

Lumber in a store is identified by its *nominal* size, which is its rough dimension before it's trimmed and sanded to its finished size at the lumber mill. The actual finished dimensions are always slightly different from the nominal dimensions. For example, what a lumberyard calls *1 inch x 6 inch lumber* is in fact ¾ inch x 5½ inch. The Material column in the following table lists nominal dimensions and the Dimensions column lists the actual, final measurements.

You can adjust the stand's height to suit your needs by adjusting the length of the 4-inch-x-4-inch cedar posts. Longer legs result in less bending over during inspections. But keep in mind that the higher the stand, the higher your honey supers will be, potentially making it more difficult to lift the heavy, honey-laden supers off the hive. I find the 13-inch height of this design just right for me.

Quantity	Material	Dimensions	Notes
4	4" x 4" cedar posts	12¼" x 3½" x 3½"	These are the leg posts of the stand. Rabbet 5½" wide by ¾" deep along one end of the post (this rabbet accommodates the narrow sides of the stand).
4	1" x 6" of cedar or knotty pine	24" x 5½" x ¾"	These are the long sides of the stand and wide struts for the top.
2	1" x 6" of cedar or knotty pine	24" x 2" x ¾"	These are the narrow struts for the top.
2	1" x 6" of cedar or knotty pine	22½" x 5½" x ¾"	These are the short sides of the stand.

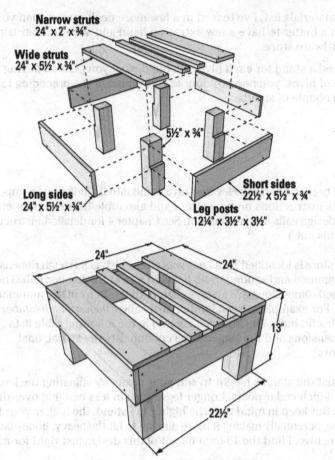

Narrow struts
24" x 2" x ¾"

Wide struts
24" x 5½" x ¾"

5½" x ¾"

Short sides
22½" x 5½" x ¾"

Leg posts
12¼" x 3½" x 3½"

Long sides
24" x 5½" x ¾"

24"

24"

13"

22½"

Illustration by Felix Freudzon, Freudzon Design

Assembling the Elevated Hive Stand

Congratulations on a good job of cutting out all the parts. Now it's time to build your "table."

1. Attach the two short sides of the stand to the leg posts.

Using deck screws and your power drill with a #2 Phillips head bit, fasten the short sides (those that measure 22½ inch x 5½ inch x ¾ inch) into the rabbet cut of the posts and secure them with two screws into each leg post. The edge of each short side rail should be flush with each post. Stagger the placement of the screws to prevent splitting the wood.

Consider using a weatherproof wood glue in addition to the screws. It helps make the hive stand as strong as possible. Apply a thin layer of glue wherever the wood parts are joined together.

2. **Attach the two long sides of the stand to the leg posts and short sides.**

   Use deck screws to secure the long sides (those that measure 24 inch x 5½ inch x ¾ inch) to the assembly. Put one screw into the leg post and another into the edge of the short side rail. Use two screws for each corner. The long sides match up with the edges of the short sides.

3. **Attach the wide and narrow struts to the top.**

   The wide and narrow struts attach to the top edges of the sides. Position the wide struts at opposite ends of the stand assembly. The top of the stand is square, so you decide where they go — it makes no difference. The struts align flush with the outer edges of the assembly. Use four deck screws per wide strut, screwed through the strut and into the top edges of the long and short sides. Now take the two narrow struts and position them centered between the wide struts. Precise placement is not at all critical. Just space them evenly by eye. Secure them using one deck screw at each end of each strut. Screws go through the struts and into the top edges of the sides. (See the figure in the earlier section "Cut List" for placement of the wide and narrow struts.)

4. **Protect the wood from the elements.**

   If you're using pine (rather than cedar), paint, varnish, or polyurethane the entire hive stand assembly. Use two or three coats, letting each coat dry completely before adding the next coat. If you elect to paint the stand, any color will do — it's up to you.

Putting your hive on the elevated stand is pretty straightforward — it sits centered on the stand. Which way you orient the hive on the stand doesn't really matter, but if you're using an IPM screened bottom board (see Chapter 14), my personal preference is to orient the hive so that as much of the bottom board as possible sits over the stand's open slats. This provides the maximum "open" space under the screened bottom board, allowing mites to fall to the ground. It also assures maximum ventilation.

You have even less expensive options for providing your bees with an elevated hive stand. How does free sound? Consider a couple of salvaged cinder blocks as a hive stand. Or level off a tree stump to get your hive off the ground. Just be sure you securely bolt your bottom board to the leveled-off top of the tree stump. These two options may not be quite as elegant as building your own hive stand, but they'll do the trick!

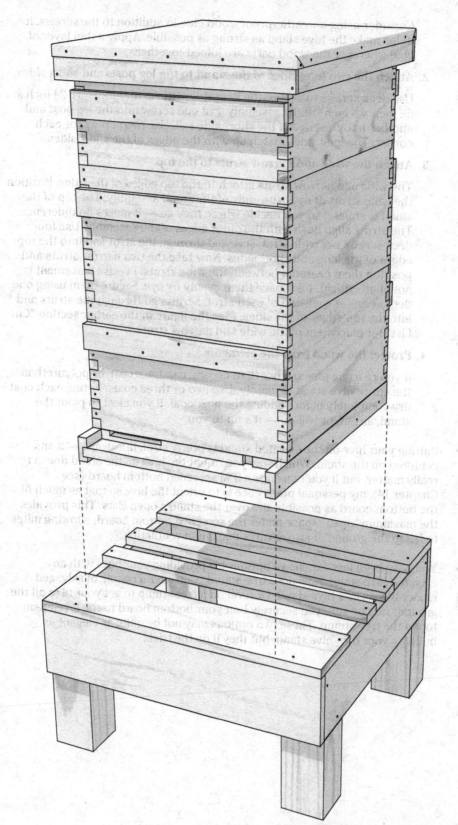

Chapter 14
The IPM Screened Bottom Board

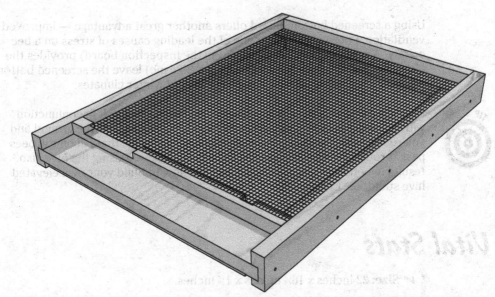

PM stands for *integrated pest management*, and the screened bottom board that I show you how to build in this chapter has become a standard part of IPM practices. The screened bottom board replaces the standard bottom board on a Langstroth hive (see Chapter 10). Its bottom is completely open, except for the screening that makes up its "floor." The unit also has a removable *inspection board* (sometimes called a *sticky board*) that you can use to monitor the colony's varroa mite population.

With varroa mites a problem for many beekeepers, screened bottom boards are gaining popularity. Here's how a screened bottom board works: A moderate number of mites naturally fall off the bees each day and land on the bottom board of the hive. Ordinarily, they would just crawl back up and reattach themselves to the bees, but not when you use a screened bottom board in place of a regular bottom board. Mites drop off the bees and either fall to the ground or are trapped on the sticky inspection board placed *under* the screening. Either way, they're unable to crawl back up into the hive. When using an inspection board, you can actually count the number of mites that have fallen off the bees and thus monitor the mite population and take action when required.

Using a screened bottom board offers another great advantage — improved ventilation. Poor ventilation is one of the leading causes of stress on a bee colony. A screened bottom board (sans the inspection board) provides the ultimate in ventilation. Some beekeepers (like me) leave the screened bottom board on the colony all year long, even in cold winter climates.

An IPM screened bottom board works best when you use it in conjunction with an elevated hive stand. When the entire hive is well off the ground and you're not using the sticky inspection board, any mites that fall off the bees pass through the screen and onto the ground below. Raising the hive also results in better air circulation. To find out how to build your own elevated hive stand, see Chapter 13.

Vital Stats

- **Size:** 22 inches x 16¼ inches x 1⅞ inches.

- **Capacity:** You need to build one screened bottom board for each hive in your apiary.

- **Universality:** The dimensions of this design are intended for use with the Langstroth hive (see Chapter 10). However, if you modify your measurements a little, a version of this will also do nicely with the nuc hive (see Chapter 6), the Warré hive (see Chapter 8), or the British National hive (see Chapter 9).

- **Degree of difficulty:** This board is likely one of the easiest projects in this book; it's no more involved than making a standard bottom board (see Chapter 8 for instructions on making a standard bottom board).

- **Cost:** As with any hive parts, using scrap wood would keep material costs minimal. But even if you purchase the recommended knotty pine, hardware, and fasteners, you can likely build a screened bottom board for less than $20.

Materials List

The following table lists what you'll use to build your IPM screened bottom board. In most cases, you can make substitutions as needed or desired. (See Chapter 3 for help choosing alternative materials.)

Lumber	Hardware	Fasteners
1 6′ length of 1″ x 5″ knotty pine lumber	1½ yards 2-ply nylon string/twine (you'll likely have to purchase a roll of this stuff)	12 #6 x 1⅝″ deck screws, galvanized, #2 Phillips drive, flat-head with coarse thread and sharp point
	⅛″ (#8) hardware cloth (typically comes in 3′ wide and 10′ long rolls, but some beekeeping supply vendors sell it by the foot)	14 5⁄32″ x 1⅛″ flat-head, diamond-point wire nails
	White Plasticor corrugated board (available at art supply stores in various sizes). Make sure you order a piece that will allow you to wind up with board that's 22″ x 14″ x 3⁄16″	24 ⅜″ staples for use in a heavy-duty staple gun
	Optional: weatherproof wood glue	
	Optional: a pint of latex or oil exterior paint (white or any light color), exterior polyurethane, or marine varnish	

Here are a few tips for purchasing materials for your IPM screened bottom board:

- ✔ I use knotty pine for my screened bottom board because it's easy to find and inexpensive. A more durable choice (but a bit more expensive) is cedar or cypress (both of which are very weather-resistant).

- ✔ In the materials list, I've tossed in a few more fasteners than you'll need just in case you lose a few along the way. It's better to have a few extras on hand and save another trip to the hardware store.

- ✔ Each hive in your apiary can use one screened bottom board. If you have a couple of hives that need boards, just double everything in the preceding table to build a couple of boards.

Cut List

This section breaks down the IPM screened bottom board into its individual components and provides instructions on how to cut those components. (*Note:* You'll make a couple of dado cuts for this bottom board. See Chapter 4 for instructions on making this cut.)

Lumber in a store is identified by its *nominal* size, which is its rough dimension before it's trimmed and sanded to its finished size at the lumber mill. The actual finished dimensions are always slightly different from the nominal dimensions. For example, what a lumberyard calls *1 inch x 5 inch lumber* is in fact ¾ inch x 4½ inch. The Material column in the following table lists nominal dimensions and the Dimensions column lists the actual, final measurements.

With some adjustment to these measurements, you can scale and modify this design to work with the Warré hive (see Chapter 8), the British National hive (see Chapter 9), or even the nuc hive (see Chapter 6). The screened bottom board's overall length and width should match the overall length and width of the standard bottom board specified for these different hive designs.

Quantity	Material	Dimensions	Notes
2	1" x 5" knotty pine	22" x 1⅛" x ¾"	These are the side rails. Dado ¾" wide by ⅜" deep along the entire length.
2	1" x 5" knotty pine	16¼" x 2" x ¾"	These are the long sides of the "floor."
2	1" x 5" knotty pine	14¾" x ¾" x ¾"	One piece is the rear cleat. The other piece is the entrance reducer (see Chapter 10 for detailed cuts).
1	1" x 5" knotty pine	15½" x 4" x ¾"	This is the wide front side of the "floor."
1	1" x 5" knotty pine	15½" x 2" x ¾"	This is the narrow rear side of the "floor."
1	Plasticor corrugated art board	22" x 14" x ³⁄₁₆"	This is the removable inspection board.
1	#8 hardware cloth	18" x 14½"	This is the screened panel that covers the opening in the floor.

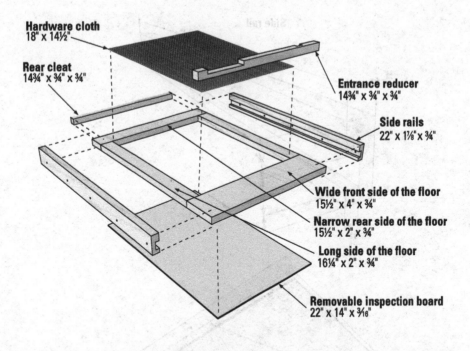

Hardware cloth
18" x 14½"

Rear cleat
14¾" x ¾" x ¾"

Entrance reducer
14¾" x ¾" x ¾"

Side rails
22" x 1⅛" x ¾"

Wide front side of the floor
15½" x 4" x ¾"

Narrow rear side of the floor
15½" x 2" x ¾"

Long side of the floor
16¼" x 2" x ¾"

Removable inspection board
22" x 14" x ³⁄₁₆"

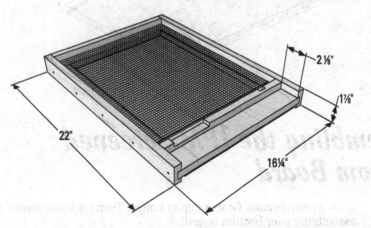

2 ⅛"

1⅛"

22"

16¼"

Illustration by Felix Freudzon, Freudzon Design

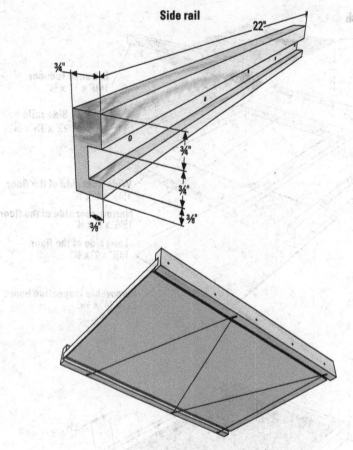

Illustration by Felix Freudzon, Freudzon Design

Assembling the IPM Screened Bottom Board

Draw a deep breath. Grab a cup of coffee. Turn on some music. And start assembling your bottom board!

1. Assemble the "floor."

Place the two long sides and the front and rear sides flat on your work surface. Essentially, you're assembling a square donut. You don't need to nail or glue anything yet.

2. **Attach the side rails and the back cleat to the floor.**

Take the two side rails and insert the "floor" assembly into the dado cut of each rail.

Be certain that the dado faces the same way in *all* rails (the dado isn't centered along the rail). Otherwise, you'll have a seriously lopsided bottom board!

Place one of the deck screws halfway into the center of each of the two side rails (the screws go through the rails and into the edges of the floor assembly). Don't screw them all the way in quite yet. First make sure everything fits properly and is square — you won't have room for adjustment after all the screws are in! When the fit looks good, use four additional deck screws spaced evenly (by eye) along each long rail. Use the first figure in the earlier section "Cut List" as a visual guide for placing the screws. These screws hold that square donut together!

The screws will go in easier if you first drill a ⅟₆₄-inch hole in each spot you plan to place a screw. The pre-drilling also helps prevent the wood from splitting.

Consider using weatherproof wood glue in addition to the deck screws. It helps make the bottom board as strong as possible. Apply a thin coat of glue wherever the wooden parts are joined together.

Nail the rear cleat to the floor assembly, using four evenly spaced flathead, diamond-point wire nails.

3. **Toe-nail the floor.**

Turn the floor over and toe-nail its four sides for additional strength (see Chapter 4 for more about the toe-nailing technique). One nail placed in each corner will do fine.

4. **Attach the screening material.**

Attach the #8 hardware cloth to the top of the "opening" of the bottom board. Use ⅜ inch staples spaced approximately 2 inches apart. Staples go around the entire perimeter of the screening. Just make sure that there are no gaps where the bees could squeeze through.

5. **Add the nylon twine to the bottom of the assembly.**

Staple the 2-ply nylon twine to the underside of the bottom board with a staple gun, using a zigzag pattern, as shown in the preceding figure. Keep the twine as taut as possible. The twine holds the removable inspection board in place.

Nylon twine has a propensity to unravel. Prevent this by singeing the cut ends of the twine with a flame.

6. **Slide the inspection board into the assembly.**

Turn the bottom board right-side-up and slide the inspection board (the Plasticor corrugated art board) into place under the screened area with the twine mesh holding it in place.

Many beekeepers (me included) put a thin coating of petroleum jelly on the top surface of the inspection board. This provides a sticky surface that ensures that any mites that fall on the board remain where they fall (like flypaper). Just wipe it off and reapply after each inspection.

7. **Place your hive on the screened bottom board.**

The screened bottom board replaces a standard bottom board. With your screened bottom board on the ground or on your hive stand (the ideal option), simply sit your Langstroth hive right on top of the screened bottom board, as shown in the following figure.

Don't paint the parts of the screened bottom board that are *inside* the hive. However, if you paint, stain, varnish, or polyurethane your hives, you're welcome to do the same to the *edges* of the bottom board — that is, those sides that are seen from the outside. This provides some protection from the elements and certainly helps the screened bottom board visually blend in with the rest of your hive.

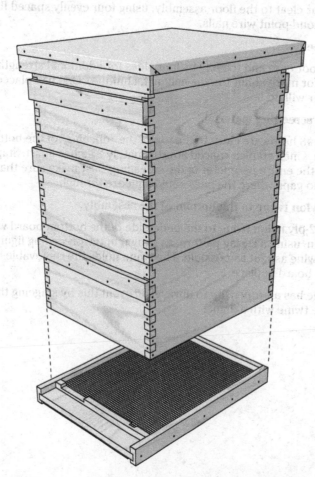

Illustration by Felix Freudzon, Freudzon Design

Chapter 15

The Hive-Top Feeder

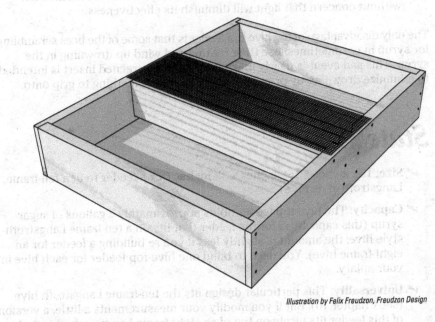

Illustration by Felix Freudzon, Freudzon Design

You use feeders to offer sugar syrup to your bees when the nectar flow is minimal or nonexistent. You'll likely need to feed your bees in the early spring and again in the autumn. Feeders also provide a convenient way to medicate your bees if you decide to do so (many medications can be dissolved in sugar syrup and fed to your bees).

Although various methods exist for feeding bees, the hive-top feeder is the one I prefer to other options. Sometimes referred to as a Miller hive-top feeder, it's based on a popular design first developed by Charles C. Miller, a commercial beekeeper during the late 19th and early 20th centuries. As the name implies, the hive-top feeder sits directly on top of the upper deep brood box and under the outer cover (you don't use an inner cover when a hive-top feeder is in place). This particular design has a reservoir that can hold a couple of gallons of syrup. Bees enter the feeder from below by means of a screened access.

The hive-top feeder has several distinct advantages over other types of feeders:

- Its large capacity means that you don't have to fill the feeder more than once every week or two.

- The screened bee access means that you can fill the feeder without risk of being stung (the bees are on the opposite side of the screen).

- Because you don't have to completely open the hive to refill it, you don't disturb the colony (every time you smoke and open a hive, you set the bees' progress back a few days).

- Because the syrup isn't exposed to the sun, you can add medication without concern that light will diminish its effectiveness.

The only disadvantage of the hive-top feeder is that some of the bees scrambling for syrup may sometimes lose their footing and wind up drowning in the syrup. This sad event is nearly inevitable, but the screened insert is intended to minimize drowning by providing the bees with something to grip onto.

Vital Stats

- **Size:** 19⅞ inches x 16¼ inches x 3½ inches (for a feeder to fit a ten-frame Langstroth-style hive).

- **Capacity:** The hive-top feeder holds approximately 2 gallons of sugar syrup (this capacity is for the feeder that fits on a ten-frame Langstroth-style hive; the amount is slightly less if you're building a feeder for an eight-frame hive). You need to build one hive-top feeder for each hive in your apiary.

- **Universality:** This particular design fits the ten-frame Langstroth hive (see Chapter 10). But if you modify your measurements a little, a version of this feeder fits nicely on top of an eight-frame Langstroth. (I include an alternative cut sheet in this chapter for a feeder that will fit the eight-frame Langstroth.) By making other adjustments to the measurements, you can tweak the feeder to fit on a Warré hive (see Chapter 8) or a British National hive (see Chapter 9).

- **Degree of difficulty:** This feeder is a little tricky to build because of the screening material and the construction detail within the feeding area. Ensuring that the feeder is watertight and doesn't leak can also be tricky. Other than that, not much more is involved than building a simple wooden box.

- **Cost:** Using scrap wood (if you can find some) would keep material costs of this design very minimal, but even if you purchase the recommended lumber, hardware, and fasteners, you can likely build this hive-top feeder for less than $50 (a little less if you use knotty pine lumber).

Materials List

The following table lists what you'll use to build your hive-top feeder. In most cases, you can make substitutions as needed or desired. (See Chapter 3 for help choosing alternative materials.)

Lumber	Hardware	Fasteners
1 12' length of 1" x 4" clear pine lumber	Optional: weatherproof wood glue	28 6 x 1⅜" deck screws, galvanized, #2 Phillips drive, flat-head with coarse thread and sharp point
1 2' x 4' sheet of ¼" lauan plywood	A pint of exterior polyurethane or marine varnish	12 ⁵⁄₃₂" x 1⅛" flat-head, diamond-point wire nails
	A tube of silicone caulking (as used for fish aquariums) or melted beeswax	⅜" staples for use in a heavy-duty staple gun
	⅛" hardware cloth (typically comes in 3' x 10' rolls, but some beekeeping supply vendors sell #8 hardware cloth by the foot; you only need a couple of feet for this project)	

Here are a few tips for purchasing materials for your hive-top feeder:

✔ I like clear pine; it's not too expensive as board lumber goes. Knotty pine is even less expensive.

✔ I've tossed in a few more fasteners than you'll need just in case you lose or bend a few along the way. It's better to have a few extras on hand.

✔ One hive uses one feeder, so if you plan on having a couple of hives, you need to double all the materials in the preceding table to build a couple of hive-top feeders.

Cut List

This section breaks down the hive-top feeder into its individual components and provides instructions on how to cut those components. I include separate cut sheets for ten- and eight-frame Langsrtoth hives. The assembly instructions for each are identical.

REMEMBER

Lumber in a store is identified by its *nominal* size, which is its rough dimension before it's trimmed and sanded to its finished size at the lumber mill. The actual finished dimensions are always slightly different from the nominal dimensions. For example, what a lumberyard calls *1 inch x 4 inch lumber* is in fact ¾ inch x 3½ inch. The Material column in the following table lists nominal dimensions and the Dimensions column lists the actual, final measurements.

TIP

This design makes use of dado and rabbet cuts. See Chapter 4 for detailed instructions on making these cuts.

Feeder for a ten-frame Langstroth hive

Quantity	Material	Dimensions	Notes
2	1" x 4" clear pine	19⅞" x 3½" x ¾"	These are the long side boards of the feeder. Cut a ¼" wide by ⅜" deep dado along the entire length of the side boards, ½" from the bottom edge. Dado two channels ¾" wide by ⅜" deep along the long dimension of the side boards. Space the dado cuts ¾" apart at the center mark of the side boards. Rabbet ¾" wide by ⅜" deep along both ends of the boards.
2	1" x 4" clear pine	15½" x 3½"x ¾"	These are the short side boards of the feeder. Cut a ¼" wide by ⅜" deep dado along the entire length of the side boards, ½" from the bottom edge.
2	1" x 4" clear pine	15½" x 2½" x ¾"	These are the shallow entrance walls to the feeding area.
2	1" x 4" clear pine	14¾" x 2⅝" x ¾"	These are the deep entrance walls in the feeding area.
2	¼" lauan plywood	15½" x 9¼" x ¼"	This becomes the "floor" of the feeder.
2	#8 hardware cloth	14¾" x 5"	These are the screened inserts for the feeding area. Fold each in half the long way, making a V shape that is 14¾" long by 2½" wide.
1	#8 hardware cloth	16¼" x 5"	This is the screened top for the feeding area.

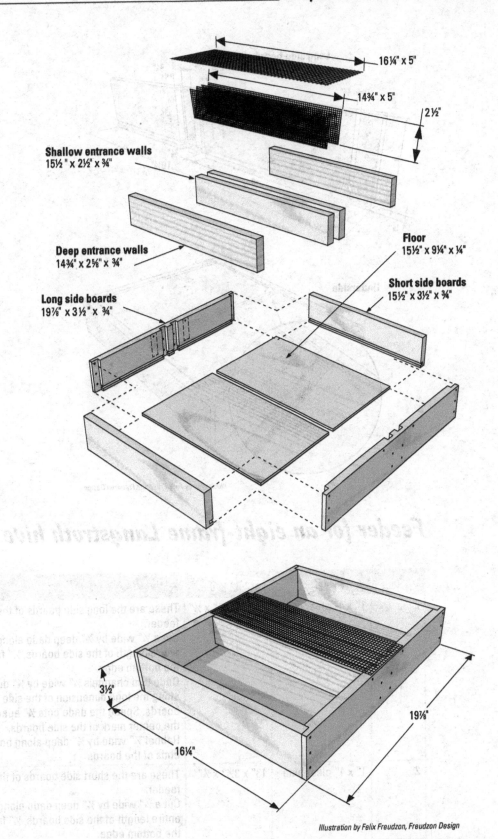

16¼" x 5"

14¾" x 5"

2½"

Shallow entrance walls
15½ " x 2½" x ¾"

Deep entrance walls
14¾" x 2⅝" x ¾"

Floor
15½" x 9¼" x ¼"

Short side boards
15½" x 3½" x ¾"

Long side boards
19⅞" x 3½" x ¾"

3½"

19⅞"

16¼"

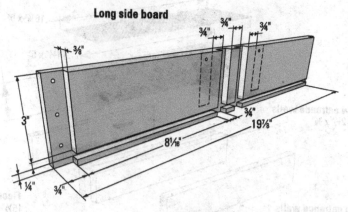

Long side board

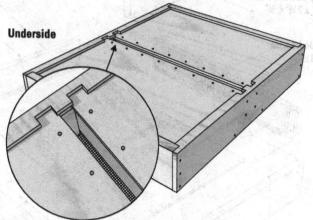

Underside

Illustration by Felix Freudzon, Freudzon Design

Feeder for an eight-frame Langstroth hive

Quantity	Material	Dimensions	Notes
2	1" x 4" clear pine	19⅞" x 3½" x ¾"	These are the long side boards of the feeder. Cut a ¼" wide by ⅜" deep dado along the entire length of the side boards, ½" from the bottom edge. Dado two channels ¾" wide by ⅜" deep along the long dimension of the side boards. Space the dado cuts ¾" apart at the center mark of the side boards. Rabbet ¾" wide by ⅜" deep along both ends of the boards.
2	1" x 4" clear pine	13" x 3½" x ¾"	These are the short side boards of the feeder. Cut a ¼" wide by ⅜" deep dado along the entire length of the side boards, ½" from the bottom edge.

Quantity	Material	Dimensions	Notes
2	1" x 4" clear pine	13" x 2½" x ¾"	These are the shallow entrance walls to the feeding area.
2	1" x 4" clear pine	12¼" x 2⅝" x ¾"	These are the deep entrance walls in the feeding area.
2	¼" lauan plywood	13" x 9¼" x ¼"	This becomes the "floor" of the feeder.
2	#8 hardware cloth	12¼" x 5"	These are the screened inserts for the feeding area. Fold each in half the long way, making a V shape that is 12¼" long by 2½" wide.
1	#8 hardware cloth	13¾" x 5"	This is the screened top for the feeding area.

Assembling the Hive-Top Feeder

With all the parts cut out and ready to go, it's time to assemble your feeder. The process is identical whether the feeder you're assembling is for a ten- or eight-frame Langstroth hive.

1. **Attach the feeder's short sides to the floor.**

   Start with one short side board. Insert one of the plywood floor pieces into the dado groove of the side board. Repeat this process with the second short side board and the second piece of plywood.

2. **Attach the feeder's long sides to the short sides and the floor.**

   Fit the rabbet edge of both long boards to the edges of the short side boards, with the dado grooves securing the edges of the plywood floor pieces. Note that there's a ¾-inch gap at the center of the floorboards. This is the bees' entrance into the feeder.

3. **Fit the shallow entrance walls into the feeder.**

   Using a hammer, tap the two shallow entrance walls into the center dado cuts of the long side boards.

4. **Screw together all the sides and the shallow entrance walls.**

   Place the assembly flat on the worktable with one of the long side boards against a "stop," such as a short piece of 2x4 clamped to the table.

   Using the deck screws and a power drill with a #2 Phillips head bit, start securing the corners of the long boards into the edges of the short side boards. Start at the end away from the "stop." Put three screws in each long side, screwing them into the edges of the two short side boards. Reverse the entire feeder end to end and screw the other corners of the long boards in a similar manner. Make certain the entire assembly remains snug and tight as you do this by using the "stop" for leverage.

   Now use a total of eight screws to secure the shallow entrance walls to the long side walls (two screws in each).

5. **Nail together the shallow entrance walls and the floor.**

Turn the feeder over (upside down) and use the wire nails to secure the edges of the plywood floor to the shallow entrance walls. Eight nails per wall should do it (see the earlier figure).

Consider using weatherproof wood glue in addition to the nails. It helps make the outer cover as strong as possible. Apply a thin coat of glue wherever two pieces of wood are joined together.

6. **Make the feeder watertight with varnish and silicone.**

Apply two or three coats of exterior polyurethane or marine varnish on the entire inside wooden surfaces of the feeder. Let each coat dry completely before applying the next coat. Pay particular attention to all the seams (where the wooden parts come together). Let the last coat dry overnight.

Apply a bead of fish tank silicone to every interior seam (anywhere the liquid might seep through). Alternatively, in place of silicone, you can use melted beeswax, sloshed into all the seams to create a watertight feeder.

Take the two deep entrance walls and coat them with two or three coats of exterior polyurethane or marine varnish. Be sure to let each coat dry completely before adding the next coat. Let the last coat dry 24 hours.

As an option, consider painting, staining, or varnishing the exterior parts of your feeder (the parts exposed to the weather) to match the look of the hive it will sit on. Doing so protects the wood from the elements.

7. **Attach the deep entrance walls to the feeder.**

Tap the two deep entrance walls into position with a hammer or rubber mallet, positioned ¾ inch from the shallow entrance walls and flush with the top of the feeder (note that this leaves a ⅜-inch gap between the bottom of these deep walls and the feeder floor; this gap allows the syrup to reach the feeding area). Secure with a total of six deck screws, screwing them through the long side boards and into the edges of the deep entrance walls.

8. **Add the hardware cloth to the feeder.**

Take the two V-shaped pieces of hardware cloth and shove them into the two spaces between the deep and shallow entrance walls. The V faces downward. These screened inserts are pressure-fitted, so you don't need to secure them with fasteners. This way you can easily remove them to clean the feeder before putting it away for the season.

Take the 16¼ inch x 5 inch piece of hardware cloth and center it on top of the feeder assembly area. Secure it with a staple gun and staples (you'll use about ten staples in all).

Don't use more staples than you absolutely need. Use just what's needed to keep the screen in place, with no gaps for the bees to squeeze through. At some point you may want to pop out these staples so that you can thoroughly clean the area within the feeder assembly.

Now that you've built your hive-top feeder, it's time to place it on the hive. It simply sits on the uppermost hive body (or super if you're using one). You don't use the inner cover while the feeder is on the hive. Fill the feeder with up to 2 gallons of sugar syrup and place the outer cover on top of the feeder. That's all there is to it — your bees are about to enjoy a banquet!

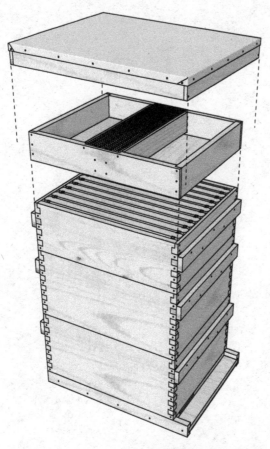

Illustration by Felix Freudzon, Freudzon Design

Here's another option for feeding your bees that costs next to nothing: a "baggie feeder"! Into a 1-gallon sealable plastic bag, pour 3 quarts of sugar syrup. Zip the baggie and lay it flat, directly on the top bars of your hive. Note the air bubble that forms along the top of the bag. Use a razor blade to make a couple of 2-inch slits into the air bubble. Squeeze the bag slightly to allow some of the syrup to come through the slits (this helps the bees "discover" the syrup). Now you need to place an empty super and outer cover on the hive (to cover the feeder).

Note that this baggie feeder is cost-effective and eliminates the danger of your bees drowning, but you do have to disrupt the bees to smoke and open the hive to put new bags on. Also, the bags aren't reusable; you have to replace them every time you add syrup.

Chapter 16

The Solar Wax Melter

Illustration by Felix Freudzon, Freudzon Design

When you extract honey from a hive, the wax cappings that you slice off represent your major wax harvest for the year. There's also the excess comb you remove during each routine inspection of a hive. Save all this wax. You'll probably get 1 or 2 pounds of wax for every 100 pounds of honey that you harvest, plus whatever burr comb you remove. You can melt down and clean this wax for all kinds of terrific uses, such as making candles, furniture polish, hand cream, lip balm, and so on. Pound for pound, wax is worth more than honey, so it's definitely worth a bit of effort to reclaim this prize and start some fun, bee-related craft projects!

The solar wax melter is a great way to render the beeswax to use for other purposes. Essentially, you melt the raw wax into a block that you can refine for various craft projects. And best of all, this device is all-natural, using no electricity — only the awesome power of the sun.

Vital Stats

- ✔ **Size:** 27¾ inches x 18⁹⁄₁₆ inches x 18½ inches.

- ✔ **Capacity:** Depending on the size of the pan you use in the melter, this design should provide ample capacity to render up to 6 to 8 pounds of wax at a time.

- ✔ **Degree of difficulty:** The butt joinery is the simplest method for assembling wood, and this design has one straightforward dado cut. All in all, an easy build. (Check out Chapter 4 for an introduction to different types of joints and cuts.)

- ✔ **Cost:** Using scrap wood (if you can find some) would keep material costs of this design minimal, but even if you purchase the recommended wood, hardware, glazing, and fasteners, you can likely build this solar wax melter for less than $75. The most expensive single item is the greenhouse glazing.

Materials List

The following table lists what you'll use to build your solar wax melter. In most cases, you can substitute other lumber as needed or desired. (See Chapter 3 for help choosing alternate materials.)

	Lumber	Hardware		Fasteners
1	10' length of 1" x 3" knotty pine lumber	A 2-pound size disposable aluminum loaf pan (approximately 8" x 4" x 2")	60	#6 x 1⅜" deck screws, galvanized, #2 Phillips drive, flat-head with coarse thread and sharp point
2	4' x 4' sheets of ¾" exterior plywood	A large, disposable aluminum roasting pan (approximately 17" x 14" x 3")	8	½₂" x 1⅛" flat-head, diamond-point wire nails
		1 2' x 4' polycarbonate dual-wall 6mm greenhouse panel (available from greenhouse supply stores and sometimes found on online auction sites like eBay)		
		A quart of flat black exterior paint (either latex or oil)		
		Optional: weatherproof wood glue		

I strongly urge you to use the recommended "plastic" (polycarbonate) glazing for the window. Not only does it work well, it also avoids all the potential dangers associated with fragile window glass. Keep in mind that this melter typically sits on the ground, and I can tell you from firsthand experience that a playful child or bouncing pet could be seriously injured stepping on a glass top. Invest in safety and use the polycarbonate greenhouse panel. Its dual-wall design is also more effective at retaining heat than a single pane of glass.

I've included a few more screws and nails than you'll use. If you're like me, you'll lose a few along the way. It's better to have a few extras on hand and save another trip to the hardware store.

Cut List

The following sections break down the solar wax melter into its individual components and provide instructions on how to cut those components.

Lumber in a store is identified by its *nominal* size, which is its rough dimension before it's trimmed and sanded to its finished size at the lumber mill. The actual finished dimensions are always slightly different from the nominal dimensions. For example, what a lumberyard calls *1 inch x 3 inch lumber* is in fact ¾ inch x 2½ inch. The Material column in the following table lists nominal dimensions and the Dimensions column lists the actual, final measurements.

The top assembly requires a dado cut. See Chapter 4 for detailed instructions on making this type of cut.

Floor assembly

Quantity	Material	Dimensions	Notes
3	1" x 3" knotty pine	6½" x 1" x ¾"	These are the retaining cleats that hold the aluminum pans in position.
1	¾" exterior plywood	18½" x 15" x ¾"	This is the upper floor.
1	¾" exterior plywood	15" x 9½" x ¾"	This is the front panel.
1	¾" exterior plywood	15" x 9¼" x ¾"	This is the lower floor.
1	¾" exterior plywood	15" x 6¾" x ¾"	This is the rear panel.
1	¾" exterior plywood	15" x 2½" x ¾"	This is the riser.

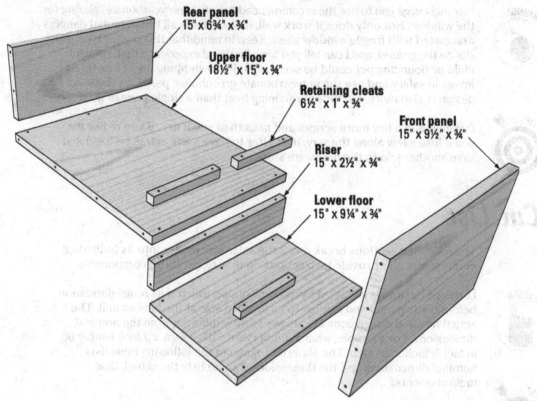

Rear panel
15" x 6¾" x ¾"

Upper floor
18½" x 15" x ¾"

Retaining cleats
6½" x 1" x ¾"

Front panel
15" x 9½" x ¾"

Riser
15" x 2½" x ¾"

Lower floor
15" x 9¼" x ¾"

Illustration by Felix Freudzon, Freudzon Design

Inclined side panels

Quantity	Material	Dimensions	Notes
2	¾" exterior plywood	27" x 17½" x 10¼" x 3/4"	These are the inclined side panels. To create an incline of approximately 15 degrees, the front edge of the panel is 10¼" high and the rear edge is 17½" high.

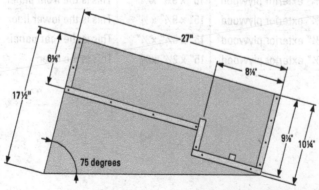

27"

6¾"

8¼"

17½"

9¼"   10¼"

75 degrees

Illustration by Felix Freudzon, Freudzon Design

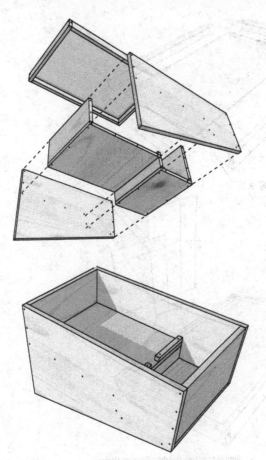

Illustration by Felix Freudzon, Freudzon Design

Glazed top assembly

Quantity	Material	Dimensions	Notes
2	1" x 3" knotty pine	27½" x 2½" x ¾"	These are the long rails of the frame. Dado a $\frac{5}{16}$" wide by $\frac{3}{8}$" deep groove along the entire length of what will be the inside of each rail. Position the bottom edge of the dado 1½" from what will be the bottom edge of the rail.
2	1" x 3" knotty pine	18½" x 2½" x ¾"	These are the short rails of the frame. Dado a $\frac{5}{16}$" wide by $\frac{3}{8}$" deep groove along the entire length of what will be the inside of each rail. Position the bottom edge of the dado 1½" from what will be the bottom edge of the rail.
1	Polycarbonate dual-wall 6mm greenhouse panel	28⅛" x 17⅝" x ¼" (6mm)	This is the window. Cut it to size using your table saw and a general purpose blade.

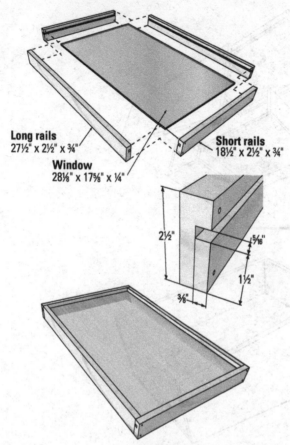

Long rails
27½" x 2½" x ¾"

Short rails
18½" x 2½" x ¾"

Window
28⅛" x 17⅝" x ¼"

2½"

⁵⁄₁₆"

1½"

⅜"

Illustration by Felix Freudzon, Freudzon Design

Assembling the Solar Wax Melter

It's time to roll up your sleeves, clear a good work space, and put all this stuff together. You start with the floor of the solar wax melter and work your way to the top.

1. Attach the riser to the lower and upper floors.

First use the deck screws and a power drill with a #2 Phillips head bit to attach the lower floor to the riser. Doing this is easiest with these elements of the floorboard assembly turned upside down on your work surface (the screws go through what will be the underside of the lower floorboard and into the lower edge of the vertical riser). This is a simple butt joint. Just line up the edges so that they're flush with each other. See the figure in the earlier section "Floor assembly" for placement.

Now flip these two components right-side up on the worktable and use deck screws to attach the upper floorboard to the top edge of the vertical riser; just line up the edges so that they're flush with each other.

Throughout the entire floor assembly, consider using a weatherproof wood glue in addition to the screws. It helps make the structure as strong as possible. Apply a thin coat of glue wherever the wooden parts are joined together (the exception is the glazed top, which you do *not* want to glue, just in case you need to replace the glazing).

The screws will go in easier if you first drill a ⁷⁄₆₄-inch hole in each spot you plan to place a screw. The pre-drilling also helps prevent the wood from splitting. Refer to the earlier figures to determine where the screws go.

2. **Attach the cleats to the upper and lower floors.**

First use the nails and a hammer to attach the two cleats to the upper floorboard. Position the cleats flush with the front and side edges of the upper floorboard. You'll have a 2-inch gap left in the center (for the melted wax to flow through). A couple of nails per cleat will do the trick. See the following figure for the approximate placement of the nails.

Now take the remaining cleat and attach it to the lower floorboard using the nails. To determine the exact placement, place your smaller disposable aluminum loaf pan on the bottom floor, as shown in the following figure. This helps you determine where to attach the cleat (the dimensions of these pans vary from brand to brand). The objective is to position this cleat so that the pan doesn't slide out of position. After all, it will be on a 15 degree incline. This pan collects the melting wax as it flows from the larger pan that sits on the upper floor. A couple of nails per cleat should the trick. See the following figure for the approximate placement of the nails.

3. **Attach the floor assembly to the front and rear vertical panels.**

Turn the entire floor assembly over and attach the front and rear vertical panels to the front and rear of the floor assembly. The large panel attaches to the lower floor, and the smaller panel attaches to the upper floor. Use deck screws to attach the panels to the floor assembly. The screws go through the lower and upper floors and into the edges of the vertical panels. These are simple butt joints. The edges of the pieces should be flush with each other. I use three screws per panel, spaced approximately as indicated in the following figure (precise placement isn't critical). Now flip the entire thing over again and proceed to the next step.

4. **Attach the side panels to the floor assembly.**

Use deck screws to attach the floor assembly (which now includes the front and rear panels) to the side panels. The screws go through the side panels and into the edges of the floor assembly. Note that the floor assembly is tilted within the side panels so that gravity will do its trick and direct the melting wax into the collection pan.

You'll make life a lot easier if you first pre-drill ⁷⁄₆₄-inch guide holes in the side panels. Lay a side panel on the workbench and then position the edge of the entire floor assembly on the side panel (just as it will go when screwed together). Use a pencil to trace the outline of the floor assembly's edges on the side panel. Do the same thing for the other side panel. Now drill the guide holes in the side panels as shown in the earlier section "Inclined side panels." This little step makes it a lot easier to correctly align and attach the floor assembly within the two side panels. Otherwise it will be a hit or miss exercise.

I use about a dozen screws on each side panel. Use the following figure to determine the approximate placement of screws. The objective is to make sure that screws go into the edges of all the critical components of the floor assembly: rear panel, upper floor, riser, lower floor, and front panel.

5. Build the glazed top assembly.

Use deck screws to attach one of the short rails to the two long rails. These are simple butt joints. You're essentially building a picture frame. Take care to align and match up the dado grooves; these are the channels into which the window panel fits. Use two screws per corner (avoid placing a screw where it may interfere with the dado groove).

Now take the polycarbonate window panel and slide it into the dado grooves of the partially assembled frame. Assuming the glazed panel was cut perfectly "square," it will square up the frame nicely.

Using two deck screws per corner, attach the remaining short rail to the long rails. Again, be careful to avoid placing a screw where it may interfere with the dado groove and the newly installed window panel.

Using screws (versus nails) allows you to remove the glazed panel at a later date, should it ever need replacing. For this reason, do *not* use wood glue on the glazed top assembly.

6. Paint all wooden surfaces matte black.

To protect the wood and better retain solar heat, paint all the wood surfaces, inside and out, using a matte black exterior paint. Two or three coats will do the trick. Let each coat dry completely before adding the next.

7. Place the aluminum pans inside the solar wax melter.

Cut a 2-inch-wide flap centered along one of the long sides of the large roasting pan (see the following figures). This hole and flap allow the melting wax to flow into the smaller pan below. The larger roasting pan sits on the inclined *upper* floor, with the cut-out flap aligned with the gap between the two retaining cleats. Fill the large pan with your wax cappings and other harvested comb.

The smaller loaf pan sits on the *lower* floor, snug against the riser and aligned to collect the melting wax from the larger pan above.

8. Put the glazed top assembly on the solar wax melter.

The removable top fits on and over the top of the wax melter (like a hat fits on a head). Position the entire unit so that the glazed top is exposed to the direct sun (facing south is best). Now all you need are some warm, sunny days and you'll soon have a lovely block of pure, natural beeswax. Time to make candles, furniture polish, and cosmetics!

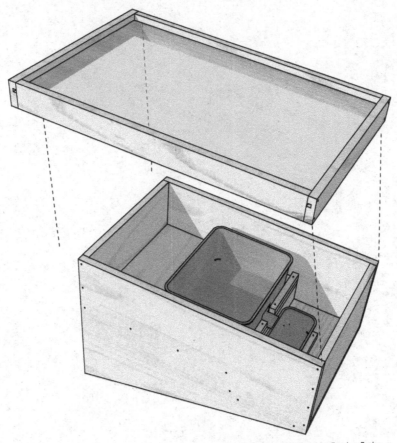

Illustration by Felix Freudzon, Freudzon Design

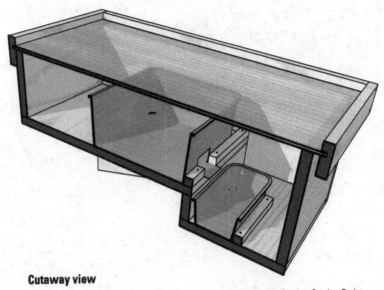

Cutaway view

Illustration by Felix Freudzon, Freudzon Design

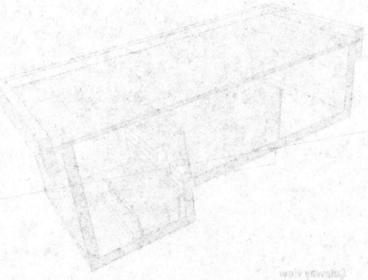

Chapter 17

Langstroth Frames

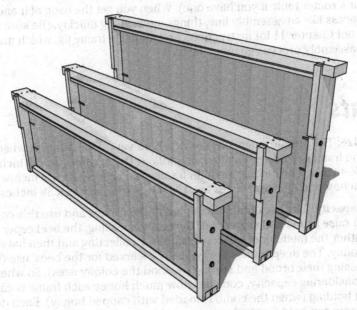

Illustration by Felix Freudzon, Freudzon Design

Three of the hives in this book use Langstroth-style frames: the five-frame nuc hive in Chapter 6, the four-frame observation hive in Chapter 7, and the Langstroth hive in Chapter 10 (both eight- and ten-frame versions). The configuration of frames is much like a picture frame: a wooden rectangle, with something displayed in the center. The frame firmly holds a sheet of wax foundation and enables you to remove the panels of honeycomb for inspection or honey extraction.

Frames come in three basic sizes — deep, medium, and shallow — corresponding to deep hive bodies and medium or shallow honey supers. The method for cutting and assembling deep, medium, and shallow frames is identical. Regardless of its size, each frame has four basic components: one top bar with a wedge, one bottom bar with a slit or groove running its length, and two side bars. The only difference between the frame sizes is the vertical measurement of the side bars. Each frame holds foundation, which consists of a thin, rectangular sheet of wax embossed with a comb pattern; it encourages your bees to draw even and uniform honeycombs. In this chapter, I include the cut lists and instructions for making deep, medium, and shallow frames, so you can build whatever size you prefer.

Making frames from scratch is complex because it involves many fussy cuts and lots of time. Most people opt to purchase frames from a beekeeping supply vendor rather than make them from scratch. They're fairly inexpensive to buy (around $1 per frame) and come as a kit that you can nail together easily. Nevertheless, some die-hards will want to take a crack at building frames from raw lumber, particularly those who relish the idea of making every single component of their hives. If that sounds like you, this chapter offers a great opportunity for you to give frame building a try. Take your time and follow the instructions carefully. You can make all the cuts with your table saw (or a router table if you have one). When you get the hang of it and set up the process like an assembly line, things move along quickly. (Be sure to check out Chapter 11 for instructions on building a frame jig, which makes the frame-assembly process go much faster.)

Vital Stats

- ✔ **Size:** The measurements of the side bars vary depending on whether the frames are *deep, medium,* or *shallow.* Deep frames are 19 inches x 1⅛₆ inches x 9⅛ inches; medium frames are 19 inches x 1⅛₆ inches x 6¼ inches; and shallow frames are 19 inches x 1⅛₆ inches x 5⅝ inches.

- ✔ **Capacity:** The bees build wax comb in the frames and use this comb to raise brood and store food. Generally speaking, the beekeeper uses either the medium or shallow frames for collecting and then harvesting honey. The deep frames are typically reserved for the bees' use (for raising their brood and storing the food the colony uses). So when considering capacity, consider how much honey each frame is capable of holding (when the comb is loaded with capped honey). Each deep frame can hold 6 pounds of extractable honey; each medium frame can hold 4 pounds of honey; and each shallow frame can hold 3 pounds of honey.

- ✔ **Universality:** These frames are designed to fit only those hives that accept the very popular and widely used Langstroth-style frames. In this book, that includes the nuc hive (Chapter 6), the observation hive (Chapter 7), and the Langstroth hive (Chapter 10).

- ✔ **Degree of difficulty:** This is likely the most intense of all the builds in this book, mostly because there are so many little details and so many parts to cut out and assemble.

- ✔ **Cost:** The materials, hardware, and fasteners to build ten frames will likely run around $15, or less if you use scrap wood.

Materials List

The following table lists what you'll need to build 10 frames of any size (deep, medium, or shallow). Multiply these quantities by 2 if you plan to build 20 frames, by 3 if you need 30 frames, and so on.

Lumber	Hardware	Fasteners
1 4' length of 1" x 8" clear pine lumber	10 sheets of crimp-wire beeswax foundation. Select the size that corresponds to the size of frame you plan to build (deep, medium, or shallow).	65 $\frac{5}{32}$" x 1⅛" flat-head, diamond-point wire nails
1 8' length of 2" x 3" spruce or fir	Optional: weatherproof wood glue	35 ⅝" finish brad nails
		45 foundation support pins

Here are a few notes about these frame materials:

✔ Be sure to save all the scrap wood from your other projects; you may just have enough on hand to build the frames you need. Waste not, want not!

✔ Beeswax foundation isn't something you're likely to make yourself; the equipment needed to make it is expensive. You can buy foundation for Langstroth-style frames from any beekeeping supply store (many suppliers are on the Internet; just do a search for "beekeeping supplies").

Note: Depending on your beekeeping supplier, crimp-wire foundation is sometimes called *hooked-wire foundation.* For this design you want the kind of foundation where the vertical wires protrude from one side and are bent at right angles, and the wires on the other side are trimmed flush with the foundation.

✔ I've tossed in a few extra fasteners in case (like me) you lose or bend a few along the way. It's better to have a few extras on hand and save yourself a trip to the hardware store.

✔ You won't find foundation pins at your local hardware store. This specialty item is available only from beekeeping supply vendors (many are on the Internet; I suggest you order the foundation pins when you order your beeswax foundation).

✔ These plans assume making ten frames at a time, but after you set up your workshop to make frames, why not crank out as many as you can? You can always use an inventory of extra frames.

Cut List

The following sections break down the frames into their individual components and provide instructions on how to cut those components for deep, medium, and shallow frames. I also provide details on making some particularly tricky cuts for side bars and top bars. Refer to the size of frame you plan to build.

Lumber in a store is identified by its *nominal* size, which is its rough dimension before it's trimmed and sanded to its finished size at the lumber mill. The actual finished dimensions are always slightly different from the nominal dimensions. For example, what a lumberyard calls *1 inch x 8 inch lumber* is in fact ¾ inch x 7½ inch. The Material column in the following tables lists nominal dimensions, and the Dimensions column lists the actual, final measurements.

Deep frames

Quantity	Material	Dimensions	Notes
20	2"x 3" spruce or fir	9⅛" x 1⅜" x ⅜"	These are the side bars. Drill two ⅛" holes 1¾" apart and centered top to bottom and left to right. These holes are for the foundation pins. Note the special cuts I describe in the later section "Making tricky cuts for side bars."
10	1"x 8" clear pine	19" x 1¹⁄₁₆" x ¾"	These are the top bars. Note the special cuts I describe in the later section "Making tricky cuts for top bars."
10	1"x 8" clear pine	17¾" x ¾" x ⅜"	These are the bottom bars. Cut a saw kerf centered along the entire length, ⅛" wide by ⁵⁄₁₆" deep.
10	Sheets of 8½" deep crimp-wire beeswax foundation		

Medium frames

Quantity	Material	Dimensions	Notes
20	2"x 3" spruce or fir	6¼" x 1⅜" x ⅜"	These are the side bars. Drill two ⅛" holes 1¾" apart and centered top to bottom and left to right. These holes are for the foundation pins. Note the special cuts I describe in the later section "Making tricky cuts for side bars."
10	1"x 8" clear pine	19" x 1¹⁄₁₆" x ¾"	These are the top bars. Note the special cuts I describe in the later section "Making tricky cuts for top bars."
10	1"x 8" clear pine	17¾" x ¾" x ⅜"	These are the bottom bars. Cut a saw kerf centered along the entire length, ⅛" wide by ⁵⁄₁₆" deep.
10	Sheets of 5⅝" medium crimp-wire beeswax foundation		

Shallow frames

Quantity	Material	Dimensions	Notes
20	2"x 3" spruce or fir	5⅜" x 1⅜" x ⅜"	These are the side bars. Drill two ⅛" holes 1¾" apart and centered top to bottom and left to right. These holes are for the foundation pins. Note the special cuts I describe in the later section "Making tricky cuts for side bars."
10	1"x 8" clear pine	19" x 1¹⁄₁₆" x ¾"	These are the top bars. Note the special cuts I describe in the later section "Making tricky cuts for top bars."
10	1"x 8" clear pine	17¾" x ¾" x ⅜"	These are the bottom bars. Cut a saw kerf centered along the entire length, ⅛" wide by ⁵⁄₁₆" deep.
10	Sheets of 4¾" shallow crimp-wire beeswax foundation		

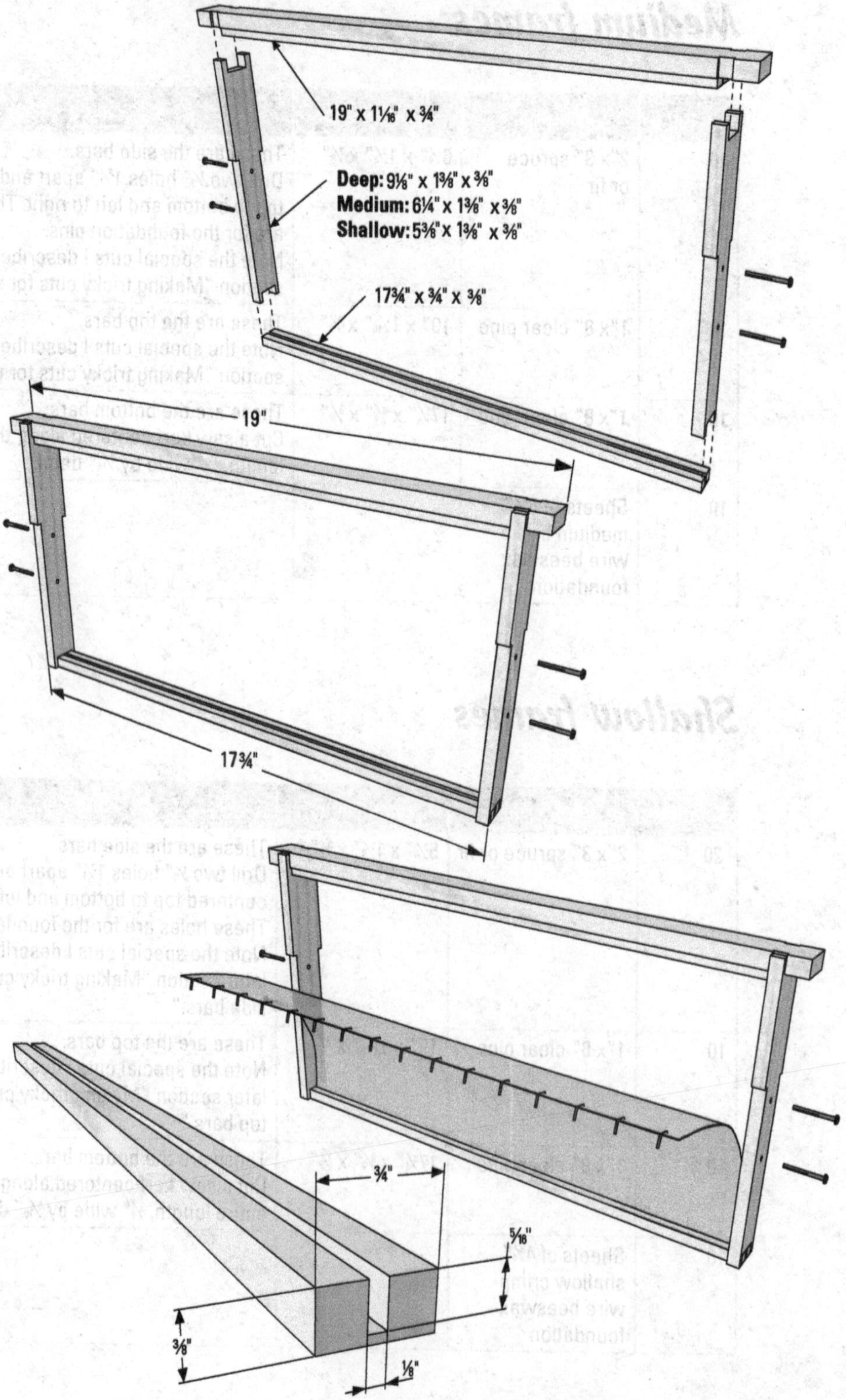
19" x 1 1/16" x 3/4"
Deep: 9 1/8" x 1 3/8" x 3/8"
Medium: 6 1/4" x 1 3/8" x 3/8"
Shallow: 5 3/8" x 1 3/8" x 3/8"
17 3/4" x 3/4" x 3/8"
19"
17 3/4"
3/4"
5/16"
3/8"
1/8"

Illustration by Felix Freudzon, Freudzon Design

Making tricky cuts for side bars

Side bars have a wide profile at the top and taper to a more narrow profile at the bottom. This tapered shape provides the correct distance between frames and allows for the proper bee space around and in between the frames so that bees can travel freely (and so they won't glue the frames together; see Chapter 4 for more about bee space). Each end bar has a notch at the top to accommodate the top bar and a notch at the bottom to accommodate the bottom bar. Follow these steps and refer to the following figure to make the cuts for side bars. The basic steps are identical for deep, medium, and shallow side bars. Make these cuts using your table saw or a table router if you have one.

I find it easiest to make frames assembly-line fashion — I set up my work space for making one particular cut and then repetitively make that cut on *all* the pieces that call for it. For example, I cut out all the top notches on all the side bars before readjusting my tools and measurements and moving on to the bottom notches.

1. **Create the taper by removing ³⁄₁₆ inch of material from each vertical edge of the bar.**

   Note that the lower portion of the side bar is more narrow than the upper portion. Refer to the call-outs on the figure to determine where to begin the taper.

2. **Cut a notch ⅛ inch wide by ⁷⁄₁₆ inch deep at the top of the bar.**

   The top bar will snap into this notch when you assemble the frame.

3. **Cut a notch ¾ inch wide by ⅜ inch deep at the bottom of the bar.**

   The bottom bar will snap into this notch when you assemble the frame.

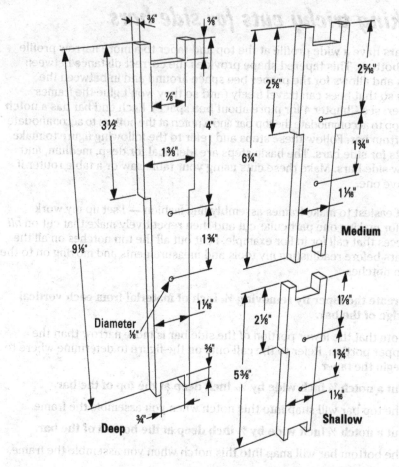

3/8"   3/8"

2⁵⁄₁₆"

7/8"

3½"   4"

1¾"

1⅜"

6¼"

1¹⁄₁₆"

Medium

9⅛"

1¾"

1¾"

Diameter ⅛"

1¹⁄₁₆"

1⅛"

2⅛"

3/8"

1¾"

5⅜"

3/4"   **Deep**

1¹⁄₁₆"

Shallow

Illustration by Felix Freudzon, Freudzon Design

Making tricky cuts for top bars

The top bars require the trickiest cuts. Follow these steps and refer to the following figure to make these cuts.

1. **Cut a kerf ⅛ inch wide by ⁵⁄₁₆ inch deep along the entire long length of the designated underside of the top bar.**

2. **Cut a vertical notch ⅜ inch wide by ³⁄₃₂ inch deep on both sides and at each end of the top bar.**

   The notch starts ⅝ inch back from the ends of the bar. When you assemble the frames, you insert the top of the side bars into these notches.

3. **On the underside of each end of the bar, make a 1 inch wide by ⅜ inch deep rabbet.**

   This creates the tabs at each end of the top bar.

4. Cut a kerf ⅛ inch wide by ⁹⁄₁₆ inch deep along the entire long length of one of the vertical sides of the top bar.

Position the cut ⅜ inch down from the top of the bar. Save the resulting strip of wood. This piece is *not* scrap — it becomes the *wedge bar* that you'll use later on when you install foundation.

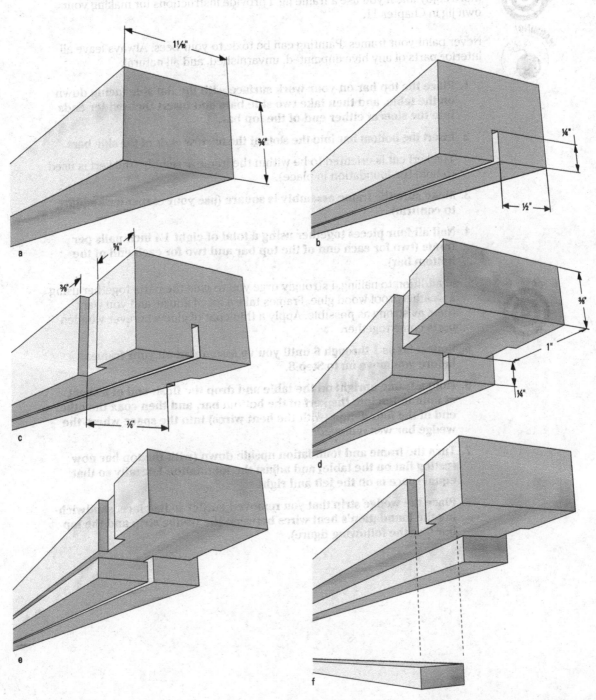

Illustration by Felix Freudzon, Freudzon Design

Assembling Langstroth Frames

You assemble all frames, no matter the size, in an identical manner. Follow these instructions for each frame you put together.

Putting together frames can be monotonous, but the assembly is faster and more enjoyable if you use a frame jig. I provide instructions for making your own jig in Chapter 11.

Never paint your frames. Painting can be toxic to your bees. Always leave all interior parts of any hive unpainted, unvarnished, and all-natural.

1. **Place the top bar on your work surface with the flat side facing down on the table, and then take two side bars and insert their wider ends into the slots at either end of the top bar.**

2. **Insert the bottom bar into the slots at the narrow ends of the side bars.**

   The kerf cut is oriented to be within the frame assembly (the kerf is used to hold the foundation in place).

3. **Make sure the frame assembly is square (use your carpenter's square to confirm).**

4. **Nail all four pieces together using a total of eight 1⅛ inch nails per frame (two for each end of the top bar and two for each end of the bottom bar).**

In addition to nailing, I strongly urge you to glue the parts together using a weatherproof wood glue. Frames take a lot of abuse, and you want them as strong as possible. Apply a thin coat of glue wherever wooden parts come together.

5. **Repeat Steps 1 through 6 until you've assembled all your frames before you move on to Step 8.**

6. **Hold a frame upright on the table and drop the flush end of a sheet of foundation into the kerf of the bottom bar, and then coax the other end of the foundation (with the bent wires) into the space where the wedge bar was removed.**

7. **Turn the frame and foundation upside down (with the top bar now resting flat on the table) and adjust the foundation laterally so that equal space is on the left and right.**

8. **Place the wedge strip that you removed earlier to its place, sandwiching the foundation's bent wires between the wedge strip and the top bar (see the following figure).**

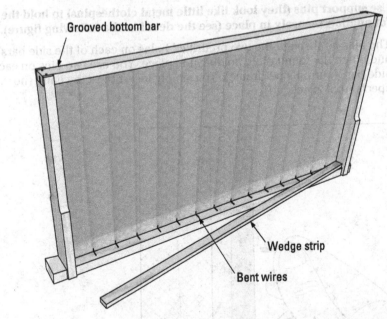

Grooved bottom bar

Wedge strip

Bent wires

Illustration by Felix Freudzon, Freudzon Design

9. **With a hammer (or better yet, a brad driver), use ⅝ inch brads to tack the wedge strip to the top bar (see the following figure).**

   Start with one brad in the center and then add one brad at each end of the wedge strip (three brads total is sufficient; you don't use glue with the wedge strip).

Illustration by Wiley, Composition Services Graphics

10. **Use support pins (they look like little metal clothespins) to hold the foundation securely in place (see the detail in the following figure).**

   The pins go through the two predrilled holes on each of the side bars and pinch the foundation, holding it in place. You use two pins on each side bar (four pins per frame). You simply insert these by hand (no special tool is needed).

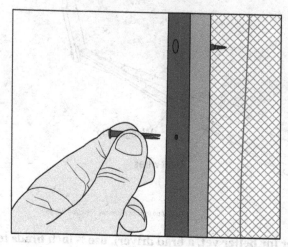

Illustration by Wiley, Composition Services Graphics

Part IV
The Part of Tens

The 5th Wave — By Rich Tennant

In this part . . .

Everyone loves a top-ten list. In this part I present three of them. The first provides practical advice for extending the life of your hives and accessories. After all, you put a lot of work into building your hives, so you may as well make them last as long as possible. The second top-ten list is all about fun and creativity — ten ways you can dress up your hives to make them more unique, more elegant, and more practical. The third list provides ten fun beehive facts to share — enjoy!

Chapter 18

Ten (Or So) Tips for Extending the Life of Your Equipment

In This Chapter

▶ Beginning with quality materials and records

▶ Inspecting, fixing, and cleaning your hives at regular intervals

▶ Elevating your hives and providing them with adequate ventilation

▶ Guarding against dangerous pests

*B*uilding your own beehives and equipment provides a great sense of satisfaction. In order to enjoy the fruits of your labor for many seasons to come, this chapter contains ten (or so) ideas for ensuring you get as much life from your new equipment as possible.

Don't Go Cheap on Materials

Saving money is always tempting, but think about the long term as you build your hives. When you compromise quality early on, you pay for it later on. Avoid the temptation. Throughout this book I suggest using galvanized nails that don't rust, outdoor quality plywood that holds up better to the elements, and weatherproof glue that doesn't deteriorate. These things may cost you a little more upfront, but they'll surely make your hives last longer. Don't cut corners. Pay for materials that will last a long time. You ultimately get what you pay for.

Keep a Build Log

Keeping notes on when you build various hive parts and components is a good idea. Referring to your build log occasionally helps you determine when it may be time to consider replacing parts.

- ✔ Replace frames and foundation every two to three years — particularly the foundation, as old wax foundation can retain contaminants and the comb can even harbor viruses that are harmful to bees. In addition, each year the size of the comb cells can get smaller, and that's detrimental to healthy brood rearing.

- ✔ Replace woodenware (hive bodies, supers, bottom boards, and so on) when rot sets in and the wood becomes spongy or soft.

- ✔ Replace woodenware when it cracks or splits, creating unwanted "extra entrances" into the hive.

- ✔ Replace roofing material (or the entire roof) when leaks are evident.

Establish an Inspection Routine

Every time you visit your colonies, you look to see how the bees are doing. Do they have enough room to expand? Is the queen present and laying eggs? Is the brood pattern good? Do the bees look healthy? All this and more is a standard part of every inspection.

But inspections are about more than just the bees. Use these opportunities to inspect your equipment as well. Do you need to address any issues? Be sure to add the following to your inspection checklist:

- ✔ What's the condition of the exterior wooden surfaces? Do they need a new paint job or a new coat of polyurethane? If so, make a note to tend to this during the winter or swap out the offending part with a fresh replacement part. (See the later section "Make Yearly Exterior Touch-Ups" for more details.)

Keep one set of extra hive parts at the ready, painted and waiting to go into action. That way you can quickly swap out fresh new parts for damaged ones without missing a beat.

- ✔ Is any of the wood damaged or rotting? If so, it's time to replace it. (See the later section "Replace Rotting Wood" for more info.)

- ✔ Is the colony growing? Does the hive look crowded? Is your honey super almost full (70 percent of the frames filled with capped honey)? If yes, it's time to build a new super and frames and get them on the hive.

- ✔ Is the colony low on food? Or do they need supplemental "nectar" to encourage wax production? Then it's time to add a hive-top feeder to the colony. Don't have one? Check Chapter 15 for plans on how to build one.

- ✔ How does the roof look? Any leaks? Keep your colonies dry by keeping all roofs in good repair. (See the later section "Repair the Roof" for the scoop.)

An inspection notebook is a great way to keep track of the status of your colonies as well as the status of your woodenware.

Prepare Your Hives for Winter

If you live in an area that experiences any degree of cold winter (where average temperatures drop below 40 degrees), you need to get your hives ready for winter. First, remove your extracted honey supers and store them inside. A garage or basement is ideal. That way they stay out of the elements.

Honey supers stored over the winter often get infested with wax moths. It happens. What a disheartening sight to discover the damage they can do to your newly built frames and foundation. You can deal with this situation in a couple of ways:

- The natural (and recommended) approach is to remove the frames and stick them in a deep freezer for at least 48 hours. You can do this in batches if you have a small freezer (or lots of frames). This freezing action kills whatever moth larvae may have taken up residence in the supers. After freezing, place the frames back in the supers and wrap everything up tight in big plastic garbage bags. Seal with twist ties so that no bugs or critters can get in. You can store the supers in a garage, basement, or wherever your spouse will permit.

- Alternatively, you can prevent wax moth damage in stored honey supers by fumigating the supers with PDB (paradichlorobenzene). For instructions on this approach see my book *Beekeeping For Dummies*, 2nd Edition (published by Wiley).

During your final hive inspection of the season, use your hive tool to scrape off any excess comb and propolis from the frames and hive bodies. Doing so makes manipulating the hive components in the spring easier and puts far less stress on these parts.

Wrap the hive in black tar paper (the kind used by roofers). Make sure that you don't cover the entrance or any ventilation holes (see Figure 18-1). The black tar paper absorbs heat from the winter sun, helps the colony better regulate temperatures during cold spells, and protects the wood from the harsh weather.

Because winters in New England can get really cold, I put a double thickness of tar paper over the top of the hive. Placing a rock on top ensures that cold winds don't lift the tar paper off. This also benefits the colony and protects your roof from the snow, ice, and rain that can deteriorate the wood.

Do Some Spring Cleaning

In the spring, perform your first spring inspection of the hives, and take the opportunity to do some spring cleaning.

If you have a hive with a bottom board (such as the nuc, Warré, British National, or Langstroth; see Chapters 6, 8, 9, and 10, respectively), remove the hive bodies to expose the hive's bottom board. Use your hive tool to scrape clean the bottom board of all debris.

Figure 18-1:
Wrapping your hive in tar paper protects both your bees and your wood-enware from the harsh winter weather.

Photograph courtesy of Howland Blackiston

I also bring a pail of warm water and a rag or sponge (but no soap or detergent) and wash down all the hive's exterior surfaces to remove dirt, grime, mildew, and any spotting left by the bees' cleansing flights.

Make Yearly Exterior Touch-Ups

As I suggest throughout this book, your hives will last much longer if you protect all exposed wooden surfaces with a good quality outdoor paint, weatherproof polyurethane, or marine varnish. That's a given, but it isn't the end of the story. Every year, check your hives in the late winter or very early spring to see whether you need to touch up the paint, polyurethane, or varnish. If it's early enough in the spring and the bees aren't yet flying, you can do these touch-ups in the field, with the bees safely clustered inside.

If the bees are already flying when you check the exterior, don't paint or re-varnish your hives in the field. The flying bees would almost certainly get stuck in the wet paint or varnish. In this case, paint a newly assembled hive part indoors and bring it outside to swap out with the part that needs attention.

I always use water-based paints because they dry much faster than oil-based products. This fast-cure feature helps ensure that bees don't get stuck in the product. Water-based polyurethanes don't seem to hold up as well to the elements as oil-based versions, so I prefer to use oil-based polyurethanes.

Repair the Roof

During any time of the year, if you see a roof in distress, you need to take action, not only to protect your bees from the elements but also to protect your woodenware.

Most of the hives in this book call for aluminum flashing on the roof. It's a great, long-lasting material for keeping the rain out and extending the life of your woodenware. You can use other materials on your hive roof to protect the wood and the bees. See Chapter 19 for some alternate roofing options that not only look great but also extend the life of all your hive parts.

Replace Rotting Wood

Sooner or later, your wood will rot. Some wood holds up better to the elements than others — namely, cedar and cypress (see Chapter 3 for more information on different woods and their durability). But when wood starts to go, there's no saving it. It's best to build a new part. Out with the old, and in with the new!

Elevate Your Hives

Wood rots much more quickly if it sits on the ground. Getting your hives off the damp ground is healthier for the bees and greatly extends the life of your woodenware. Here are a few ways to elevate your hives:

- Use a couple of cinder blocks under the hive (see Figure 18-2a).

- Use a leveled-off tree stump to lift your hive off the ground (see Figure 18-2b). *Note:* Be sure to bolt or screw the bottom board onto the stump so that the hive doesn't tumble off.

- A very attractive solution is to build an elevated hive stand. You can find plans for such a stand in Chapter 13.

Figure 18-2: You can use cinder blocks or a level tree stump to elevate your hives off the damp ground.

a

b

Photographs courtesy of Howland Blackiston

Provide Proper Ventilation

Proper ventilation is critical to maintaining the comfort and health of your bees. Good circulation also keeps moisture from building up inside the hive, compromising the longevity of your woodenware. You have numerous ways to ensure good ventilation:

- Use an elevated hive stand (see Chapter 13).
- Use a ventilated inner cover (see Chapters 6 and 10 for hive plans that use inner covers).
- Use a double screened inner cover for added ventilation (see Chapter 12).
- Use a screened bottom board in place of a conventional bottom board (see Chapter 14).
- Provide ventilation holes (see Chapters 5, 6, 7, 9, and 10 for hive plans that use ventilation holes).

A useful way to ventilate a hive is to drill wine cork–sized holes in hive bodies and supers (see Figure 18-3). Insert or remove the cork to control air circulation as needed.

Figure 18-3: Holes the size of wine corks are useful for hive ventilation.

Photograph courtesy of Howland Blackiston

Guard against Bears with an Electric Fence

Do bears like honey? Indeed they do! And they simply crave the sweet bee brood. (I've never tried it myself, but I suspect it's sweet.) If bears are active in your area (they're in a number of areas within the United States), taking steps to protect your hive from these lumbering marauders is a necessity.

If they catch a whiff of your hive, they can do spectacular and heartbreaking damage, smashing apart the hive and scattering frames and supers far and wide. What a tragedy to lose your bees and hives in such a violent way. Worse yet, you can be certain that after a bear discovers your hives, he'll be back, hoping for a second helping and destroying all your hard work.

 The only really effective defense against these huge beasts is installing an electric fence around your apiary. Anything short of this just won't work. Your state or local conservation department has information on installing such a fence. Some institutions may even provide financial assistance for the installation.

Chapter 19

Ten (Or So) Ways to Trick Out Your Hives

- -

In This Chapter

▶ Adding practical and decorative hardware

▶ Turning any hive into an observation hive

▶ Surveying alternative wood and roofing options

▶ Tapping into technology

- -

*A*ll you artists, eccentrics, and nonconformists out there are going to love this chapter. I fall into one or more of those categories, as I've always found pleasure in dressing, pimping out, or otherwise dolling up my hives to look different from the run-of-the-mill beehives out there. So without further ado, here are some fun ways to bring a little bling to your bees. I've organized the tips in this chapter into several categories:

✔ **Get handy with hardware.** Any interior designer would agree that changing out hardware is a quick and easy way to improve and update the appearance of a piece of furniture, kitchen cabinets, or your home generally. The same is true with beehives. The plans in this book prescribe the basics regarding fasteners and hardware, and the basics are perfectly functional and cost-effective. But for those of you who want to dare to be different, this chapter has a few embellishment ideas to whet your appetite.

✔ **Have artistic adventures.** The sky's the limit when it comes to thinking of ideas for dressing up your hives. There's no right or wrong. Just express yourself and have fun. Some ideas to inspire include using exotic woods, painting, shingling, and including a front porch.

✔ **Make any hive an observation hive.** You can use a couple of different methods to make virtually any hive an observation hive — namely, you can make your inner cover transparent and cut a window into a hive body. It's fun to see what's going on without having to smoke and open up the hive.

✔ **Try some fancy roofing options.** Have some fun with your hive's roof or outer cover. Consider such options as using unusual roof materials and making architectural alterations.

✔ **Install a webcam.** Use a little technology to keep your eyes on what's going on in your bee yard. WiFi webcams are now very cost-effective and allow you to log on to the Internet and observe your bees from anywhere in the world.

Use Decorative Handles and Embellishments

One of the easiest things you can do to dress up your hive is to swap out the wooden handles for some interesting, decorative hardware. The choices are nearly endless. A home design center or big-box hardware store is a good place to start looking. Also consider looking at marine supply stores, as they have a good selection of heavy-duty, weatherproof hardware. Decide first whether you're partial to the old-fashioned antique look (in which case a home design center would be a good bet) or the sleek and modern look (stainless steel nautical hardware may fit the bill).

Don't favor design over safety. Make sure the handles you choose are up to the task. The handles on a fully loaded super or hive body have to be able to lift between 50 and 90 pounds. You don't want the pretty hardware you choose to fail you. Dropping heavy woodenware filled with thousands of startled bees is not a good thing. Use good judgment. Affix handles with *at least* four heavy-duty screws (see Figure 19-1).

Figure 19-1: These heavy-duty flush-mount chest handles provide an excellent grip for lifting the hives in my bee yard.

Photograph courtesy of Howland Blackiston

You can find a wide selection of decorative hardware out there, particularly by perusing kitchen hardware. For example, I found interesting bee skep drawer pulls at a home goods store that I used as "awnings" over ventilation holes (see Figure 19-2). A nice decorative touch.

Here's another example: Being an amateur artist, I sculpted a bee-motif medallion in clay, made a silicone mold of the clay model, and then cast copies of this design using bronze resin. The medallions measure about 4 inches across. I typically mount one on each of my hives as a sort of "brand" (see Figure 19-3). When you have a mold, you can make as many duplicates as you want. If you're not up to the task of making one of these, a copy of this same design is available from www.bee-commerce.com.

Figure 19-2: I used bee skep drawer pulls as little awnings over ventilation holes that I drilled in some hive bodies.

Photograph courtesy of Howland Blackiston

Figure 19-3: I "brand" my hives with this faux bronze bee medallion.

Photograph courtesy of Howland Blackiston

Add Metal Frame Rests

A metal frame rest isn't so much decorative as it is practical. Wooden frame rests take a bit of a beating as you pry frames apart and scrape off burr comb and propolis during hive inspections. Over time, the soft wood gets chewed up pretty badly, and sliding frames along the frame rests becomes more difficult. The solution is to tack a narrow strip of aluminum flashing over your wooden frame rests. The metal lasts a long time, makes removing propolis easier, and protects the frame rests from getting nicked up. Do this as a final step when you're assembling your hive bodies and supers.

The length of the metal strips should be equal to the length of your frame rests, and the width should be double the measurement of the frame rest "ledge." Bend aluminum flashing at sharp right angles along the entire length — simply fold the flashing by hand over the lip of the frame rest. Secure the flashing strip using six to eight ⅝-inch brads (see Figure 19-4). Position the brads only on the vertical surface so the horizontal ledge stays smooth.

Figure 19-4:
Adding a narrow strip of aluminum flashing along the frame rests makes it easier to manipulate frames and slide them into place.

Photograph courtesy of Howland Blackiston

Employ Exotic Woods

Throughout this book I recommend lumber that's readily available, cost-effective, and appropriate for the job at hand. Chapter 3 includes information about deciding on your various wood options. But that's not what I'm talking about in this chapter. I'm talking over-the-top, exotic wood that will transform your beehives and equipment from simple to spectacular!

Quite honestly, given the cost of some woods, this may not be all that practical for day-to-day beekeeping, but wouldn't a custom-made curly maple nuc make a wonderful gift for that special beekeeper in your life? Or how about a cherry wood frame jig to present to the outgoing president of your bee club? I've made a few Langstroth hives out of African mahogany and have sold them to beekeepers with deep pockets.

Consider the observation hive (which I describe in Chapter 7). It's intended to be viewed and admired like a piece of fine furniture. For mine, I applied a veneer of book-matched burl walnut. It looks like the interior of a Rolls-Royce Phantom and draws as many oohs and aahs as the bees inside (see Figure 19-5). Just have fun and experiment.

You surely won't paint over any wood that's so beautiful and expensive, so be certain to protect your investment by applying a few coats of quality exterior polyurethane or marine varnish. Apply this *after* you have assembled the woodenware.

Figure 19-5:
I veneered this observation hive using book-matched European burl walnut, coated with a half dozen coats of high-gloss polyurethane.

Paint Creatively

People love to paint their mailboxes in creative ways, so why not their beehives? Whimsical, romantic, or abstract, there are no rules here. Just do whatever tickles your fancy. My neighbor and fellow beekeeper Marina Marchese is an artist, and she had a merry time with a couple of her helpers painting a nuc hive (see Figure 19-6). You can use any kind of exterior grade latex or oil paint. Apply your designs *after* assembling the woodenware.

Figure 19-6:
Marina Marchese's assistants Taylor Gillespie (left) and Rachel Williams (right) putting the finishing touches on an assembled nuc hive.

Shingle the Sides

Here's a fun way to make your hive look like a little cottage: After you assemble your hive bodies and supers, clad the outer sides of your hive bodies with shake shingles or traditional beveled lap siding. Attach using ⅞ inch galvanized nails. When you do this, you don't even need handles on the hive, as the lip of the siding creates a finger grip on all the hive's outer sides. Be generous with the nails (I apply them every 2 inches).

I used lap siding to create the hive shown in Figure 19-7. I purchased standard, 6-inch wide, cedar lap siding and trimmed an inch off the entire length of the thick edge (the reduced 5-inch width looks more proportional to the hive body). I finished off the hive's corners with a 1-inch-x-½-inch trim, affixing it with ⅞ inch galvanized nails spaced every few inches. I added a peaked copper-clad roof and one of my signature bronze bee medallions (refer to Figure 19-3). For fun, you can paint the hive to match the colors of your house (the siding in Figure 19-7 matches that of my own house). There you have it: a cozy cottage for your beloved bees.

TIP

The siding makes the overall outer dimensions of the hive bodies ½ inch thicker all around. So if you want your bottom board, inner cover, and outer cover to match up perfectly, add an inch to the overall dimensions of these parts when you make them.

Figure 19-7: Lap siding on standard Langstroth hive bodies.

Photograph courtesy of Howland Blackiston

Add a Front Porch

When I was in Italy, I noticed that most of the hives had a little roof just above the hive's entry. A front porch. I have no idea whether the bees prefer this little detail, but I love the way it looks. There's no reason not to add this little architectural embellishment to any hive. It may even be something that your bees do appreciate (after all, those guard bees may be grateful to get out of the rain). Add the porch *after* you assemble the hive body. I use a couple of 1¼ inch deck screws to affix the porch assembly to the front of the hive.

The example in Figure 19-8 is placed on a Langstroth hive, but by simply adjusting the overall width to match the width of your hive's entry, you can apply this front porch to most of the hive designs in this book. Now that you have a front porch, I wonder what we can do about building little rocking chairs?

Figure 19-8:
A simple front porch adds some whimsy to your hives.

Photograph courtesy of Howland Blackiston

Make Your Inner Cover Transparent

In place of the plywood insert on your inner cover or crown board (see Chapters 6, 9, and 10), use a sheet of ⅛-inch thick Plexiglas, cut to the same specified dimensions as the plywood (see Figure 19-9). This way, all you need to do to have a peek at your bees is remove the outer cover. Instant observation hive!

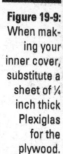

Figure 19-9: When making your inner cover, substitute a sheet of ¼ inch thick Plexiglas for the plywood.

Photograph courtesy of Howland Blackiston

Cut an Observation Window in the Hive Body

You can make almost any hive into an observation hive by cutting a window into a side of the hive body. You'll need to insert a glass or Plexiglas window and a hinged or removal panel (for the bees' privacy and to insulate the hive when you're not observing). The size of the window is up to you — it's more of an aesthetic decision than anything else. The lovely Warré observation hive in Figure 19-10 was built by the House of Bees in Southworth, Washington.

Figure 19-10: An observation Warré hive from the House of Bees.

Photograph courtesy of Darren Gordon, House of Bees

Use Alternate Roof Materials

Many of the hives in this book specify aluminum flashing for a roof material. Using copper flashing (instead of aluminum) results in a beautiful blue patina over time (like the Statue of Liberty). You apply copper flashing the same way as the aluminum flashing. Alternatively, for hives with a peak roof, such as the Kenya top bar hive (Chapter 5) or the Warré hive (Chapter 8), you can use conventional asphalt roof shingles or even cedar shake roofing. In these cases, use ⅞ inch roofing nails to attach the shingles or shakes to the roof.

Make Architectural Alterations to Your Roof

If the hive design specifies a flat roof (such as the nuc hive in Chapter 6, the British National in Chapter 9, or the Langstroth in Chapter 10), you can swap out these tops for a peaked roof (such as on the Kenya hive in Chapter 5 or the Warré hive in Chapter 8). Use the construction plans for the latter hives and simply adjust the measurement to accommodate your nuc, Langstroth, or British National hive. Just make sure the outer dimensions of the top (the part that sits on the hive) are the same as prescribed in the plans (so the roof can sit on the hive like a hat sits on a head). The lap-paneled hive shown in Figure 19-7 is a Langstroth hive with a peaked roof and copper top (and one of my bronze bee medallions).

Mount a Webcam to Your Hive

Okay, you're at work or on vacation and you miss your girls. I've been there too. But technology has provided a way to visit your bees from afar. For around $100, you can get a weatherproof, wireless webcam and mount it near your hive's entrance. By logging on to the web, you can see what your bees are up to in real time. How cool is that?

Make sure your hive and wireless camera are close enough to the house to pick up your wireless signal. Also, you need to run an electrical line to power the webcam (if you need help, consult the basic instructions that accompany your webcam).

Use Alternate Roof Materials

Many of the hives in this book specify aluminum flashing for a roof material. Using copper flashing (instead of aluminum) results in a beautiful blue patina over time (like the Statue of Liberty). You apply copper flashing the same way as the aluminum flashing, but relatively, for hives with a peak roof, such as the Kenya top bar hive (Chapter 5) or the Warré hive (Chapter 3), you can use conventional asphalt roof shingles or even cedar shake roofing. In these cases, use 3/4-inch roofing nails to attach the shingles or shakes to the roof.

Make Architectural Alterations to Your Roof

If the design specifies a flat roof (such as the one I use in Chapter 6, the British Warré in Chapter 3, or the Langstroth in Chapter 10), you can swap out those tops for a peaked roof (such as on the Kenya hive in Chapter 5 or the Warré hive in Chapter 3). Use the construction plans for the hive roof you want and simply adjust the measurement to accommodate your hive. Langstroth or British National hive, just make sure the outer dimensions of the top (the part that sits on the hive) are the same as provided in the plans (so the roof can sit on the hive like a hat sits on a head). The lap-paneled hive shown in Figure 7-1 is a Langstroth hive with a peaked roof and copper top (and one of my bronze bee medallions).

Mount a Webcam to Your Hive

Okay, you're at work or on vacation and you miss your girls. I've been there (too). But technology has provided a way to visit your bees from afar. For around $100, you can get a weatherproof, wireless webcam and mount it near your hive's entrance. By logging on to the web, you can see what your bees are up to in real time. How cool is that?

Make sure your hive and wifi/glass camera are close enough to the house to pick up your wireless signal. Also, you need to run an electrical line to power the webcam (if you need help, consult the basic instructions that accompany your webcam).

Chapter 20

Ten Fun Facts about Beehives

When I get into a new interest or hobby, I just love to get my hands on as much information as possible. I have a merry time wallowing in the details and filling my brain with fun facts and trivia. And it's a good thing, because whenever I'm at a gathering and people learn I'm into bees and beehives, they bombard me with questions. So I look like quite the one in-the-know when I retort with entertaining facts and trivia about the subjects I love.

So here's your turn. In this chapter I include some beehive-related facts that are sure to impress your colleagues in your bee club and around the water cooler. Read on, and have fun.

Discovering the First Recorded Depiction of a Beehive

The rock paintings in the Cuevas de la Araña (Spider Caves) of Valencia, Spain, are a popular tourist destination. These caves were used by prehistoric people who painted images on the stone walls of activities that were a critical part of their everyday life, such as goat hunting and, lo and behold, honey harvesting. One of the paintings depicts two men climbing up vines and collecting honey from a wild beehive. Immortalized on the rock wall between 6,000 and 8,000 years BC, this painting is widely regarded as the earliest recorded depiction of a beehive of any kind. So, looking ahead, make sure you take some digital images of yourself building the beehives in this book. You never know who may be admiring those pictures 8,000 years from now!

Unearthing the World's Oldest Beehives

Ancient Egyptian tomb and temple paintings prove that humankind has been making and tending to beehives for many thousands of years and that honey was an important commodity. But this is merely pictorial evidence of man-made beehives. Where are those old hives today? In 2007 a remarkable find was made during an archaeological dig in the Beth Shean Valley of Israel. An entire apiary was uncovered from biblical times, containing more than 30 mostly intact clay hives, some of which still held very, very old bee carcasses. These man-made hives date from the tenth to early ninth centuries BC, making them the oldest beehives in the world.

Recounting a Brief History of Beehives around the Globe

For centuries, hives have been built as containers in which to raise bees and collect honey. Around the world, hives have been made of any suitable material that was readily available in a particular area. The result has been a plethora of hive shapes, sizes, and designs — enough to fill an entire book. Here are a few examples:

- The big breakthrough in modern hive design didn't come until 1853, when Reverend Lorenzo Lorraine Langstroth published a design for the hive that still bears his name. His allowance for bee space (see Chapter 4) and removable frames revolutionized beekeeping. There have been no innovations of this magnitude since then. Flip to Chapter 10 for plans on building your own Langstroth hive and Chapter 17 for details on building Langstroth-style frames.

- In Mediterranean countries and the Middle East, hives of yore were typically fashioned from pottery or clay, shaped into long hollow cylinders, and closed at both ends, with a small hole on one end for the bees and a larger one at the other end so that the beekeeper could remove chunks of honeycomb. Containing no frames, the bees would build a free-form comb in the hive.

- In Africa, bark was stripped from trees and reassembled as long hollow pipes. The ends were filled in (leaving an entry for the bees). Alternatively, logs were chiseled out to form a cavity for the bees. These horizontal bark or log hives were then strung high in tree branches, out of the reach of predators. This horizontal style of hive ultimately evolved into top bar hives that could be easily and cheaply manufactured from available scrap wood. The bees hang their free-form comb on the removable bars placed along the entire top of the hive. The Kenya style of top bar hive has sloped sides to discourage the bees from attaching the comb to the side walls, thus making the bars of comb removable. This primitive style of hive (still widely used in Africa) is gaining renewed interest among beekeepers in developed countries. Take a journey to Chapter 5 and find out how to build your own Kenya top bar hive.

- In early Britain and Western Europe, hives were fabricated from woven baskets made of grass, straw, reeds, or bramble stems. The material was woven into a continuous wreath and spirally piled-up and laced together to form a domed basket. Sometimes the hives were then coated with clay and dung to render them more weatherproof. This picturesque style of hive is known as a *skep,* and it remained widely in use until Reverend Langstroth's more efficient design set beekeeping abuzz.

✔ Today, skeps are often used as romantic renderings of a beehive. It was this shape of hive that gave the beehive hairdo its name (for details, see the later section "Using Beehives for Design Inspiration"). However, in most developed countries, keeping bees in skeps is illegal, as they're impractical when it comes to inspecting a colony for disease. You can still purchase skeps from some beekeeping supply shops and basket stores, but these are sold for decorative use only. They do look pretty nestled into a corner of the garden.

For a truly entertaining, wonderful, deep dive into the long history of beekeeping and beehives, get a copy of *The Archaeology of Beekeeping* by Eva Crane (Gerald Duckworth and Co., Ltd.).

Bee-ing the Beehive State

In the United States, the state of Utah is lovingly known as the "beehive state." Most of the nicknames associated with Utah are related to the members of the Church of Jesus Christ of Latter-day Saints (Mormons) who first settled in the territory. Deseret, a term found in The Book of Mormon and the name Mormons proposed as their provisional state, actually means *honeybee*. The early Mormon settlers carried beehives with them to the region. In 1847 the beehive was adopted as an official emblem representing the qualities of industry, perseverance, thrift, stability, and self-reliance — all virtues respected by the Mormon faith. Today the beehive is the centerpiece of the Utah state flag and the Utah state seal. For an image of the state seal, see www.pioneer. utah.gov/research/utah_symbols/flag.html.

Studying Beehives in Outer Space

In 1984, NASA's space shuttle Challenger carried a student experiment that involved a beehive filled with honeybees. The shuttle traveled hundreds of miles above earth in near-zero gravity, and the idea was to see whether the bees could make honeycomb during the seven-day journey while weightless. It turns out that they could, and they did. Just like the astronauts, the astro-bees were, well, as busy as bees during space travel. The original Buzz Lightyear!

Finding the Largest Beehive in the World

The world's largest hive is not a man-made one; it's a living tree — a banyan tree to be specific. To find this hive, you have to travel to the village of Nandagudi in the state of Karnataka in India. The tree is home to around 600 colonies of bees, making it the largest beehive on earth. The banyan tree is surrounded by eucalyptus trees whose flowers are a major source of nectar to the resident bees. It's become quite the tourist attraction. Do an online image search for Nandagudi banyan tree and see it without having to make the journey.

Using Beehives for Design Inspiration

Beehives are amazing structures, so it's no surprise that they provide design inspiration. Here are a couple of examples:

- **Beehive hairdo:** Margaret Vinci Heldt ran a hair salon in Chicago from the 1950s through the 1970s. In 1960, *Modern Beauty Salon* magazine asked Margaret to develop a fashionable new hairstyle that would reflect the coming decade. By combing, teasing, and using lots of hair spray, her new hairstyle design achieved dramatic height. In a 2012 TV interview with Margaret, she said the piled-up style was inspired by a tall fez hat she owned and loved. The fez had decorations on it that looked like bees. As she was completing the prototype hairdo on a pretty model in a photographer's studio, she added one of the bee decorations to the new hairstyle. The representative from the magazine commented, "This looks just like a beehive!" The kitschy "big hair" beehive hairdo was born, and the style caught on like wildfire. In fact, Audrey Hepburn wore her hair in the new beehive style as she starred in the classic movie *Breakfast at Tiffany's* in 1961. Later, in recognition of this honey of an idea, Cosmetologists Chicago, a trade association with 60,000 members, created a scholarship fund in Margaret's name for her creativity in hairdressing.

- **The Beehive Houses of Italy:** In the Puglia region of Italy, you'll find many 20-foot-diameter stone buildings, called *trulli houses.* The locals have been building these beehive-inspired structures for the past 5,000 years. Originally, the beehive structures were built as tool sheds and barns, but in more recent centuries, several were built in clusters and used as homes. The beehive houses of Puglia are a notable example of mortarless construction, made entirely of limestone, with thick, solid walls and conical, stone-stacked roofs. One single keystone holds the entire structure together. The little town of Alberobello has almost 1,500 of these unique houses. If you're in the area, some of these homes are available for vacation rental.

Creating Beehives for Bumblebees

In nature, the fat, furry bumblebee lives in small colonies, usually within a grassy clump of grass, in an abandoned underground rodent's nest, under a porch, or in any convenient dry spot. Like honeybees, bumblebee colonies have a queen, worker bees, and drones (see Chapter 1 for details on these types of bees). Bumblebees collect pollen and make small quantities of honey for their food. A colony of bumblebees doesn't over-winter like honeybees, however. At the end of the season, all but a newly mated queen perishes. The queen goes underground to hibernate for the winter, to emerge again in the spring and start laying eggs that will become her new colony.

It's possible to provide man-made hives for bumblebees. They're the primary pollinators for clover, alfalfa, field beans, tomatoes, cotton, apples, plums, raspberries, and sunflowers. In fact, colonies of bumblebees are available commercially for the pollination of greenhouse tomatoes! Caring for bumblebees is different from caring for honeybees, though. For more information on this charming creature, and to download plans for building your own bumblebee nest, go to www.bumblebee.org.

Moving a Beehive without Confusing Your Bees

There's a trick to moving a beehive. If you want to move a colony of bees from the front of your house to the back of your house (say several hundred feet), it's not a matter of just hauling the beehive from point A to point B. That's because bees imprint on their local area very precisely. So when you move a hive only a short distance (less than 2 miles), the flying bees will head back to the original site (where did my beehive go?). They'll become totally lost and will certainly die.

If you intend to move the hive more than 2 miles, the bees won't recognize any of the new surrounding area, and they'll quickly learn their new location. So that kind of long-distance move isn't problematic. But if you need to move your hive less than 2 miles (and that certainly includes from the front to the back of the house), then you need to follow one of these three options:

- ✔ Move the beehive to a new site that's at least 2 miles away. Leave it there for a couple of weeks, and then return it to the desired location.

- ✔ Move the beehive 2 or 3 feet each day until you've reached the desired location. Making the move taking baby steps doesn't befuddle the bees.

- ✔ If you live in a northern area where the bees will cluster and bunker down for cold months, wait until early winter and then move the beehive to the desired location.

Transporting Migratory Beehives

Migratory beekeepers make their living transporting beehives to different geographic locations to produce specialty honey and to collect fees for pollinating fruit trees, almonds, alfalfa, and other crops. They travel around the country, following the bloom, and then head south in the fall and winter to maintain their colonies in the warmer southern weather. The United States has approximately 1,000 migratory beekeepers, transporting approximately 2 million beehives a year. They use large, 18-wheeler, flatbed trailer trucks that can each accommodate more than 500 beehives, containing nearly 20 million bees. It makes for spectacular headlines whenever one of these trucks takes a spill and releases its girls in a neighborhood.

The American Beekeeping Federation is a society of professional beekeepers. For more information on migratory beekeeping and this society, visit www. abfnet.org/index.cfm.

Moving a Beehive without Confusing Your Bees

There's a trick to moving a beehive. If you want to move a colony of bees from the front of your house to the back of your house (say several hundred feet), it's not a matter of just hauling the beehive from point A to point B. That's because bees imprint on their local area very precisely, so when you move a hive only a short distance (less than 2 miles), the flying bees will head back to the original site (where did my beehive go?). They'll become totally lost and will certainly die.

If you intend to move the hive more than 2 miles, the bees won't recognize any of the new surrounding area, and they'll quickly learn their new location. So that kind of long distance move isn't problematic. But if you need to move your hive less than 2 miles (and that certainly includes from the front to the back of the house), then you need to follow one of these three options:

- Move the beehive to a new site that's at least 3 miles away. Leave it there for a couple of weeks, and then return it to the desired location.

- Move the beehive 2 or 3 feet each day until you've reached the desired location. Making this move taking baby steps doesn't befuddle the bees.

- If you live in a northern area where the bees will cluster and hunker down for cold months, wait until early winter and then move the beehive to the desired location.

Transporting Migratory Beehives

Migratory beekeepers make their living transporting beehives to different geographic locations to produce specialty honey and to collect fees for pollinating fruit trees, almonds, alfalfa, and other crops. They travel around the country, following the bloom and head south in the fall and winter to maintain their colonies in the warmer southern weather. The United States has approximately 1,000 migratory beekeepers, transporting approximately 2 million beehives a year. They use large, 18-wheeler, flatbed tractor trucks that can each accommodate more than 500 beehives, containing upwards of 20 million bees. It makes for spectacular headlines whenever one of these trucks takes a spill and releases its guts to a neighborhood.

The American Beekeeping Federation is a society of professional beekeepers. For more information on migratory beekeeping and this society, visit www.abfnet.org/default.htm.

Math & Science

Algebra I For Dummies,
2nd Edition
978-0-470-55964-2

Biology For Dummies,
2nd Edition
978-0-470-59875-7

Chemistry For Dummies,
2nd Edition
978-1-1180-0730-3

Geometry For Dummies,
2nd Edition
978-0-470-08946-0

Pre-Algebra Essentials
For Dummies
978-0-470-61838-7

Microsoft Office

Excel 2010 For Dummies
978-0-470-48953-6

Office 2010 All-in-One
For Dummies
978-0-470-49748-7

Office 2011 for Mac
For Dummies
978-0-470-87869-9

Word 2010
For Dummies
978-0-470-48772-3

Music

Guitar For Dummies,
2nd Edition
978-0-7645-9904-0

Clarinet For Dummies
978-0-470-58477-4

iPod & iTunes
For Dummies,
9th Edition
978-1-118-13060-5

Pets

Cats For Dummies,
2nd Edition
978-0-7645-5275-5

Dogs All-in One
For Dummies
978-0470-52978-2

Saltwater Aquariums
For Dummies
978-0-470-06805-2

Religion & Inspiration

The Bible For Dummies
978-0-7645-5296-0

Catholicism
For Dummies,
2nd Edition
978-1-118-07778-8

Spirituality
For Dummies,
2nd Edition
978-0-470-19142-2

Self-Help & Relationships

Happiness
For Dummies
978-0-470-28171-0

Overcoming Anxiety
For Dummies,
2nd Edition
978-0-470-57441-6

Seniors

Crosswords For Seniors
For Dummies
978-0-470-49157-7

iPad 2 For Seniors
For Dummies,
3rd Edition
978-1-118-03827-7

Laptops & Tablets
For Seniors For Dummies,
2nd Edition
978-1-118-09596-6

Smartphones & Tablets

BlackBerry For Dummies,
5th Edition
978-1-118-10035-6

Droid X2 For Dummies
978-1-118-14864-8

HTC ThunderBolt
For Dummies
978-1-118-07601-9

MOTOROLA XOOM
For Dummies
978-1-118-08835-7

Sports

Basketball For Dummies,
3rd Edition
978-1-118-07374-2

Football For Dummies,
2nd Edition
978-1-118-01261-1

Golf For Dummies,
4th Edition
978-0-470-88279-5

Test Prep

ACT For Dummies,
5th Edition
978-1-118-01259-8

ASVAB For Dummies,
3rd Edition
978-0-470-63760-9

The GRE Test
For Dummies, 7th Edition
978-0-470-00919-2

Police Officer Exam
For Dummies
978-0-470-88724-0

Series 7 Exam
For Dummies
978-0-470-09932-2

Web Development

HTML, CSS, & XHTML
For Dummies,
7th Edition
978-0-470-91659-9

Drupal For Dummies,
2nd Edition
978-1-118-08348-2

Windows 7

Windows 7
For Dummies
978-0-470-49743-2

Windows 7
For Dummies,
Book + DVD Bundle
978-0-470-52398-8

Windows 7 All-in-One
For Dummies
978-0-470-48763-1

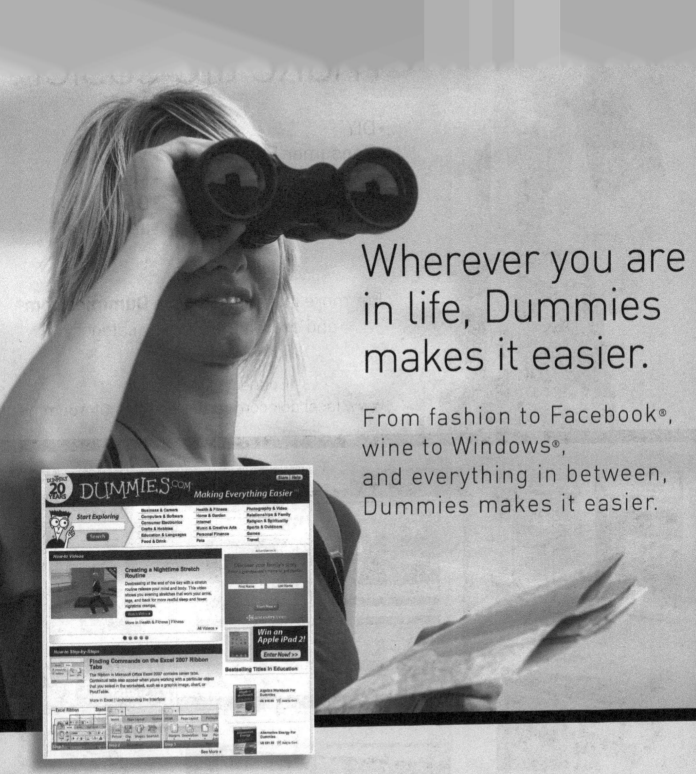

Wherever you are in life, Dummies makes it easier.

From fashion to Facebook®, wine to Windows®, and everything in between, Dummies makes it easier.